# Studies in Systems, Decision and Control

Volume 526

The series "Studies in Systems, Decision and Control" (SSDC) covers both new developments and advances, as well as the state of the art, in the various areas of broadly perceived systems, decision making and control–quickly, up to date and with a high quality. The intent is to cover the theory, applications, and perspectives on the state of the art and future developments relevant to systems, decision making, control, complex processes and related areas, as embedded in the fields of engineering, computer science, physics, economics, social and life sciences, as well as the paradigms and methodologies behind them. The series contains monographs, textbooks, lecture notes and edited volumes in systems, decision making and control spanning the areas of Cyber-Physical Systems, Autonomous Systems, Sensor Networks, Control Systems, Energy Systems, Automotive Systems, Biological Systems, Vehicular Networking and Connected Vehicles, Aerospace Systems, Automation, Manufacturing, Smart Grids, Nonlinear Systems, Power Systems, Robotics, Social Systems, Economic Systems and other. Of particular value to both the contributors and the readership are the short publication timeframe and the world-wide distribution and exposure which enable both a wide and rapid dissemination of research output.

Indexed by SCOPUS, DBLP, WTI Frankfurt eG, zbMATH, SCImago.

All books published in the series are submitted for consideration in Web of Science.

Ludwik Czaja

# Cellular Cause-Effect Structures

Springer

Ludwik Czaja
Department of Computer Engineering
Vistula University
Warsaw, Poland

ISSN 2198-4182 ISSN 2198-4190 (electronic)
Studies in Systems, Decision and Control
ISBN 978-3-031-55589-3 ISBN 978-3-031-55590-9 (eBook)
https://doi.org/10.1007/978-3-031-55590-9



This Springer imprint is published by the registered company Springer Nature Switzerland AG
The registered company address is: Gewerbestrasse 11, 6330 Cham, Switzerland

Paper in this product is recyclable.

# Preface

The existence of various descriptions of real dynamic phenomena leads to considerations on their suitability for specific applications, formal coherence or the ease and convenience of drawing conclusions, but also on the common features of various approaches. This suggests generalizing them or expressing one description in terms of another by simulation or by retaining structural features and modifying semantical ones or vice versa. A number of such attempts have been undertaken in theoretical computer science, for system models like CSP,[1] CCS,[2] CPS,[3] TR[4] DL,[5] PN,[6] SS,[7] RS,[8] CA,[9] CES.[10] This book deals with the use of the network construct of cause-effect (c-e) structures together with the accompanying algebraic operations in order to develop their variant that would behave like cellular automata. That is, which preserves the structural properties of c-e structures, but with semantics, i.e. the next-state rules, specific to a particular cellular c-e structure, as in the case of cellular automata, rather than common to all of them, as in the case of general c-e structures. Structurally, the graphs of cellular automata are a special case of c-e structures' graphs. This is due to the special connection of cells, i.e. vertices in cellular automata: each vertex is connected by bidirectional arrows to the same number of vertices called its neighbours, while the connection in the general c-e structures is arbitrary. Both system models are based on graphs, where, in general c-e structures, the predecessors of each vertex are grouped into its cause regions and its successors

---

[1] Communicating Sequential Processes, [5].
[2] Calculus of Communicating Systems, [12, 13].
[3] Concurrent Program Schemes [9].
[4] Trace Theory [10, 11].
[5] Dynamic Logic, [15, 8].
[6] Petri nets [14, 16].
[7] Substitution Systems [4].
[8] Reaction Systems, [3, 7].
[9] Cellular Automata [17, 6, 18].
[10] Cause-Effect Structures [1, 2].

into its effect regions, while in cellular automata such kind of grouping is absent.[11] A certain structural similarity of c-e structures and cellular automata is also manifested in their infiniteness: according to the original definitions both are graphs with infinite sets of vertices. All this prompts an attempt to incorporate to cellular c-e structures, some structural features of general c-e structures, and some semantical features specific for cellular automata. For instance, combining cellular c-e structures by means of summation—an operation on c-e structures, applied not only to assemble small parts into large entirety, but also to shape them into manifolds, such as cylinders or toruses, an action called here "bending the planar bases". The multidimensional cellular c-e structures, the feature of cellular automata, are considered in Chap. 4. Moreover, the shape of cells—vertices of graphs in graphical presentation—is taken into account here too. This raises the question of which polygons or polyhedra can be cells, if their sides or faces are supposed to be in the one-to-one relationship with neighbouring cells, in particular, to be adjacent to the sides or faces of neighbouring cells. So what cellular c-e structures can tile ("tessellate") space, i.e. cover it without overlapping of cells-tiles with no gaps. That is why some topics of the book pertain to geometric and topological properties of cellular c-e structures. Many pictorial presentations are designed to make these considerations visible.

Spaces containing geometric and graph-oriented objects are equipped with a coordinate system, Cartesian in this case, which enables the numerical representation of the objects' properties and their mutual relations. In particular, to express two main structural features: the equinumerous neighbourhood of all cells and a fixed order and orientation of their neighbours in space. And from the semantic side, suggests to define the state of cellular c-e structures as a function with numerical values, and the rules of the next state as relations between states. Such relations are specific to individual cellular c-e structures, not the same to all: no single rule shared by all cellular c-e structures exists, as is the case with general c-e structures.[12] The next-state rules of cellular c-e structures are represented by the so-called semantic tables in the form of matrices. Such table determines a succeeding value to be assumed by a cell depending on its current value and current values of its neighbours. The structure of semantic tables is the following. Each column represents one neighbour of a cell under examination and each row contains fixed values of its neighbours.[13] A procedure which examines the table looks for a row with entries identical to the values assumed by the neighbours of the cell in the current state. If the search is successful, the procedure decides what value the cell will take in the next state. Otherwise, the

---

[11] One may say that each region consists of single vertex. The graph-like structure of both system models regards the ordinary, not bipartite graphs (with two kinds of vertices) as in the case of Petri nets.

[12] So, only the interconnection of nodes and initial state defines behaviour of a given general c-e structure. Whereas in the cellular ones, where the interconnection is similar (isomorphic) for each node in a given cellular c-e structure, the behaviour is defined by its individual next-state rule and initial state.

[13] This means that these values are constant regardless of state and are identical for each cell, due to the same number of neighbours of all cells and the same order and orientation of their neighbours in space.

value of the cell will not change. This action is performed simultaneously for all cells.[14] The question here is about the type of relations defining the rules of the next state. For instance, is it to be a function (bijection or a function of another type), what makes deterministic choice of the next state, or a relation being not a function. It is the semantic table that encodes this binary relation specific to a task modelled by the cellular c-e structure and constructed to implement its behaviour intended by the designer. A number of examples of such tables for concrete tasks modelled by cellular c-e structures are presented in the book. In the case of a functional next-state relation, they are also pictured by ordinary sequential deterministic finite state automata in the form of multigraphs. Such multigraph visualizes behaviour of a single cell in the cellular c-e structure.

Creating cellular c-e structures by summation of smaller structures raises also questions about correctness of the result. And both: structural correctness (i.e. whether the result is a cellular c-e structure) and semantical (i.e. whether it would behave in accordance with intention of the users of this system model). Much attention is paid to this issue in the book, and especially to the correctness of constructing the semantic table for the result, created from tables for the components. These considerations end with stating criteria for the structurally and semantically correct summation of cellular c-e structures. As pointed out in Chap. 1, the problem with correctness of the result of combining elementary ordinary c-e structures does not exist.[15] This is so because there are no restrictions to interconnection of nodes as in their cellular variant and, unlike for that variant, the next-state rule is the same for all ordinary c-e structures. But especially the creation of non-planar cellular c-e structures, like cylindric or toroidal, by cut-out and combine pieces of infinite flat surfaces, makes the problem of the results' correctness vital. This is the issue of Chaps. 3, 6 and 7.

The presentation of the book's subject matter proceeds from illustrative examples of cellular c-e structures to the perception of general problems related to such model of a dynamic system and its description and discussion. The description, presumably rigorous enough to be properly understood, avoids being overly formalized. So, apart from Chap. 1, which collects necessary formal definitions and properties of general c-e structures, not the cellular, the succeeding chapters contain description of the latter largely in a plain language, but with accompanying formal substantiations, wherever a vagueness might occur. In that case, the description refers to the axioms 2.1 in Chap. 2 characterizing the commutative and idempotent monoid of cellular c-e structures, or to commonly known geometric or topological and set-theoretical concepts and properties. No special new notions or technically sophisticated terminology for the cellular c-e structures are introduced. Nor for analysing their diverse possible behaviours, entailed from generality of relations for the next-state rules. This is usually quite tedious work, bringing conjectures rather than formally demonstrated

[14] The parallel semantics is defined in Chap. 1 also for the general c-e structures.

[15] But the structural correctness of combining extended c-e structures requires the coherence of arrow weights and capacity of nodes in the components being combined. This is mentioned in Chap. 1. Lacking of this problem concerns elementary c-e structures, not the extended.

results. It can be said that the approach is more phenomenological than analytical. Similar attempt and approach has been undertaken in [2] for development of a variant of cause-effect structures that would behave like reaction systems [3]. In that case also the structural features of c-e structures have been exploited, but with semantics adopted from the reaction systems. The structural similarity of ordinary c-e structures to their reactional variant was there even closer than to their cellular variant, because both algebraic operations, addition and multiplication, were used to combine reactional c-e structures. So, all features of quasi-semiring of c-e structures have been used, not only summation, as in case of their cellular variant.

Warsaw, Poland
July 2023

Ludwik Czaja

## References


1. Czaja L.: Cause-Effect Structures. An Algebra of Nets with Examples of Application, Lecture Notes in Networks and Systems 45, Springer (2019)
2. Czaja L.: Computing in Cause-Effect, Structures, Springer (2022) (Lecture Notes in Networks and Systems 331)
3. Ehrenfeucht, A, Petre I., Rozenberg G.: Reaction Systems: A Model of Computation Inspired by the Functioning of the Living Cell, in: The Role of Theory in Computer Science, Essays Dedicated to Janusz Brzozowski, edited by Stavros Konstantinidis, Nelma Moreira, Rogério Reis and Jeffrey Shallit, World Scientific 2017, pp.1-32
4. Genrich H., Lautenbach J. K., Thiagarajan P.S.: Substitution systems — A family of system models based on concurrency, Mathematical Foundations of Computer Science 1980 pp 698–723
5. Hoare C.A.R.: Communicating Sequential Processes, Prentice Hall (1985)
6. Kari J. J.: Cellular Automata, in Handbook of Natural Computing, vol. 4, Springer (2012)
7. Kleijn J., Kountny M., Rozenberg G.: Modelling Reaction Systems with Petri Nets, in: Proc. of the International Workshop on Biological Processes & Petri Nets (BioPPN 2011), pp. 36–52
8. Kozen D., Tiuryn J.:, Logics of Programs, Chap. 14 in Handbook of Theoretical Computer Science, vol.B, Elsevier (1990)
9. Mazurkiewicz, A.: Concurrent Program Schemes and their Interpretation. DAIMI PB-78, Aarhus University, Aarhus (1977)
10. Mazurkiewicz, A.: Trace theory, in Petri nets, Applications and Relationship to other Models of Concurrency, no. 255 in Lecture Notes in Computer Science, pp. 279-324, Berlin-Heidelberg-New York, Springer (1987)
11. Mazurkiewicz, A.: Introduction to Trace Theory, in The Book of Traces, World Scientific, Singapore-New York-London-Hong Kong (1995)
12. Milner R.: A calculus of communicating systems, Springer Lecture Notes in Computer Science, 92, (1980)
13. Milner R.: Communication and Concurrency, Prentice Hall Inc. 1989
14. Petri C.A.: Communication with Automata, Final report, Vol. 1, Supplement 1, RADC TR-65-337-vol1-suppl 1, Applied Research, Princeton, NJ. Contract AF 30(602)-3324 (January 1966)
15. Pratt V.: Semantical Considerations on Floyd-Hoare Logic, Proc. 17th Annual IEEE Symposium on Foundations of Computer Science, 1976, 109-121.
16. Reisig W.: Petri Nets. An Introduction, Number 4 in EATCS Monographs on Theoretical Computer Science, Springer, Berlin-Heidelberg-New York, Tokyo, 1985

17. Wolfram S.: Universality and complexity in cellular automata. Physica D 10, 1–35, 1984
18. Zaitsev D.: Simulating Cellular Automata by Infinite Petri Nets, in Journal of Cellular Automata, January 2018

# Acknowledgements

The initiative to write the book about a version of cause-effect structures acting similarly to cellular automata belongs to Prof. Andrzej Skowron. I would like to thank him and Prof. Janusz Kacprzyk, editor of the series "Studies in Systems, Decision and Control", for promoting the book in the publishing process. A special recognition deserves Krzysztof my son, whose contribution during the work on this book was indispensable. This included much discussion on the subject of the book, his construction of computerized tools to verify samples of cellular cause-effect structures, converting the text from a non-standard LaTeX into a standard, notably giving the book proper editorial shape.

Warsaw, Poland
July 2023

Ludwik Czaja

# Contents

# About the Author

**Ludwik Czaja** obtained his M.Sc., Ph.D. and habilitation degrees from the University of Warsaw, where he was a Professor of Informatics at the Faculty of Mathematics, Informatics and Mechanics. He spent several years at other universities, like Carnegie Mellon, Oxford, Ben-Gurion, Humboldt, as a visiting professor or research fellow. Currently, he is a Full Professor at the Vistula University and an Emeritus Professor of the University of Warsaw. His work encompasses formal and programming languages, compilers, theory of computation, parallel and distributed processing.

# Chapter 1
# ABC of Cause-Effect Structures

This chapter comprises selected concepts of cause-effect (c-e) structures [2, 5, 6], useful or indispensable to go through the succeeding part of the book. The main subject of the following chapters is an attempt to exploit these concepts for presentation of a c-e structures' variant, a model of dynamic system, whose objects in structure and work, resemble the cellular automata [9, 11, 19]. A choice of the concepts is prompted by some constructs taken from the cellular automata, as both systems are rooted in the network structure. Although most of them are general mathematical notions, introduction of a few specific for the subject are justified in comments that aim at making plausible their adequacy as a modelling tool for variety of tasks, and to substantiate their formal correctness. Some of their consequences, possibly aidful in farther reading, are quoted without formal proofs from the references above mentioned. But formal definitions of necessary notions concerning c-e structures are provided. The comments sometimes refer to Petri nets, as commonly known model of dynamic system based on akin conceptions. Out of two methods of c-e structures description: an axiomatic [5, 6] and fixpoint [2]–by recursive equations, the axiomatic manner has been adopted to devise the cellular cause-effect structures. So, let us begin the basic concepts of c-e structures in their axiomatic frame.

## 1.1 Axiomatic Framework

### *1.1.1 A*

In this description, the c-e structures constitute an algebraic calculus referred to as a *quasi-semiring*. "Quasi" because of restricted axiom of distributivity of multiplication over addition, contrarily to ordinary algebraic semirings. First, the main notions and properties of elementary c-e structures are given, then some their specific problems and extensions. The formal definitions are the following:



L. Czaja, *Cellular Cause-Effect Structures*, Studies in Systems, Decision and Control 526,
https://doi.org/10.1007/978-3-031-55590-9_1

**Definition 1.1** (*set* $\boldsymbol{F}[\mathbb{X}]$*; quasi-semiring of formal polynomials*) Let $\mathbb{X}$ be a non-empty enumerable set. Its elements, called *nodes*, are counterparts of places in Petri nets. $\theta \notin \mathbb{X}$ is a symbol called *neutral*. It plays role of neutral element for operations on terms, expressions called *formal polynomials over* $\mathbb{X}$. The names of nodes, symbol $\theta$, operators $+$ ("addition") and $\bullet$ ("multiplication") and parentheses are symbols out of which formal polynomials are formed. A node name and $\theta$ is a formal polynomial; if $K$ and $L$ are formal polynomials, then terms $(K + L)$ and $(K \bullet L)$ are too. Their set is denoted by $\boldsymbol{F}[\mathbb{X}]$. Assume that multiplication binds expressions stronger than addition. Hence some parentheses may be omitted. The syntax of addition and multiplication of formal polynomials is defined as follows: $K \oplus L = (K + L)$, $K \odot L = (K \bullet L)$. To spare notation, signs $+$ and $\bullet$ instead of $\oplus$ and $\odot$ will be used. It is required that the system $\langle \boldsymbol{F}[\mathbb{X}], +, \bullet, \theta\rangle$ obeys the following equality axioms for all $K, L, M \in \boldsymbol{F}[\mathbb{X}]$, $x \in \mathbb{X}$:

| | | | |
|---|---|---|---|
| (+) | $\theta + K = K + \theta = K$ | (•) | $\theta \bullet K = K \bullet \theta = K$ |
| (++) | $K + K = K$ | (••) | $x \bullet x = x$ |
| (+++) | $K + L = L + K$ | (•••) | $K \bullet L = L \bullet K$ |
| (++++) | $K + (L + M) = (K + L) + M$ | (••••) | $K \bullet (L \bullet M) = (K \bullet L) \bullet M$ |
| (+•) | If $L \neq \theta \Leftrightarrow M \neq \theta$ then $K \bullet (L + M) = K \bullet L + K \bullet M$ | | |

The algebraic system based on these axioms is called a *quasi-semiring of formal polynomials*.[1] □

**Definition 1.2** (*cause-effect structure; carrier; set* $\boldsymbol{CE}[\mathbb{X}]$) A *cause-effect structure* (*c-e structure*) over $\mathbb{X}$ is a pair $U = \langle C, E\rangle$ of total functions:

$C : \mathbb{X} \to \boldsymbol{F}[\mathbb{X}]$ (*cause function*; nodes occurring in $C(x)$ are *causes* of $x$),
$E : \mathbb{X} \to \boldsymbol{F}[\mathbb{X}]$ (*effect function*; nodes occurring in $E(x)$ are *effects* of $x$),

such that $x$ occurs in the formal polynomial $C(y)$ iff $y$ occurs in $E(x)$. *Carrier* of $U$ is the set $car(U) = \{x \in \mathbb{X} : C(x) \neq \theta \vee E(x) \neq \theta\}$. $U$ is of *finite carrier* iff $|car(U)| < \infty$ ($|\ldots|$ denotes cardinality). The set of all c-e structures over $\mathbb{X}$ is denoted by $\boldsymbol{CE}[\mathbb{X}]$. Since $\mathbb{X}$ is fixed, we write just $\boldsymbol{CE}$. □

Let us comment the definitions in the following remarks and illustrate them by examples.

1. The converse implication to that in axiom (+•) is not satisfied. Indeed, if $K \bullet (L + M) = K \bullet L + K \bullet M$ for each $K, L, M$ then $x \bullet (\theta + (x + y)) = x \bullet \theta + x \bullet (x + y)$, thus the equivalence $L \neq \theta \Leftrightarrow M \neq \theta$ is false for $L = \theta$, $M = x + y \neq \theta$.

[1] This algebra, a case of the "algebra of terms" [13], induces a quotient algebra of equivalence classes of formal polynomials equivalent with respect to axioms (+)–(+•). It reminds the Lindenbaum-Tarski algebra if the operators "+", "•" were treated as logical connectives "∨", "∧" which, in such perception, led to the fix-point framework of c-e structures mentioned in Sect. 1.2 but is misleading in the present section.

2. Axioms (+) and (•) state that $\theta$ is the neutral for both operators. If idempotence of multiplication held for all terms, not only for the single symbols as stated in axiom (••), and if distributivity of • over + held unconditionally, then the two operations would become identical. Indeed, in this case $K + L = (K + L) \bullet (K + L) = K + L + K \bullet L$, for $K, L \in \boldsymbol{F}[\mathbb{X}]$, but $L + K \bullet L = \theta \bullet L + K \bullet L = (\theta + K) \bullet L = K \bullet L$ and similarly $K + K \bullet L = K \bullet L$. Therefore $K + L = K \bullet L$. This would make this algebraic system inadequate for building cause-effect structures as descriptive tool, alternative and equivalent to Petri nets. In fact, such algebraic system becomes a commutative and idempotent monoid.
3. Axiom (++) is equivalent to $\theta + \theta = \theta$. Indeed, (++) $\Rightarrow \theta + \theta = \theta$ follows from (+), and $\theta + \theta = \theta \Rightarrow$ (++) holds since $K = K \bullet \theta = K \bullet (\theta + \theta) = K \bullet \theta + K \bullet \theta = K + K$.
4. The axiomatic system $\langle \boldsymbol{F}[\mathbb{X}], +, \bullet, \theta \rangle$ has a model defined as follows: $\theta$ is mapped onto {}–the empty family of subsets of $\mathbb{X}$; every $x \in \mathbb{X}$ is mapped onto the family $\{\{x\}\}$ containing only the subset $\{x\}$ of $\mathbb{X}$; if $K$ and $L$ are mapped onto families $G$ and $H$ of subsets of $\mathbb{X}$, then $L + K$ is mapped onto family $G \cup H$ and $K \bullet L$ is mapped onto family:

$$M = \begin{cases} \{g \cup h : g \in G \wedge h \in H\} & \text{if } G \neq \{\} \wedge H \neq \{\} \\ G & \text{if } H = \{\} \\ H & \text{if } G = \{\} \end{cases}$$

   In this way, each formal polynomial over $\mathbb{X}$ is a notation for a certain finite family of finite subsets of $\mathbb{X}$ and conversely: each such family can be denoted by a certain formal polynomial. For instance, $a + (b + d) \bullet c$ denotes family $\{\{a\}, \{b, c\}, \{c, d\}\}$ of subsets of $\mathbb{X} = \{a, b, c, d, \ldots\}$ (Fig. 1.3). We use "{}" to denote the empty family of subsets of $\mathbb{X}$, whereas "{{}}" to denote family containing only the empty subset of $\mathbb{X}$. Existence of such model ensures consistency of the above axiomatic system and is a basis for interpretation of c-e structures as dynamic objects, alternative but behaviourally equivalent to Petri nets.
5. Intended meaning of addition and multiplication is exclusive choice and simultaneity respectively. Thus, for instance, the subscripted node $a_{x+y}$ symbolises the choice between sending token from $a$ to $x$ or to $y$, whereas $a_{x \bullet y}$–sending token from $a$ to $x$ and to $y$ simultaneously. So, c-e structures and respective Petri nets in Fig. 1.1 should act identically.
   Anticipating formal definitions of operations on c-e structures, the c-e structure at the left side in Fig. 1.1 is the sum and at the right side – product of arrows $\{a_x^\theta, x_\theta^a\}$ and $\{a_y^\theta, y_\theta^a\}$. For $y = \theta$, two possible meanings (in terms of Petri nets) of adding the neutral symbol are in Fig. 1.2.
   There are two reasons to adopt meaning expressed by the left side of Fig. 1.2. (1) Nets with "dangling" transitions have no direct equivalents amongst c-e structures; nonetheless their behaviour equivalent to that of Petri nets, may be simulated and symbolised pictorially by a zigzag arrow, as at the right side of Fig. 1.2. (2) The

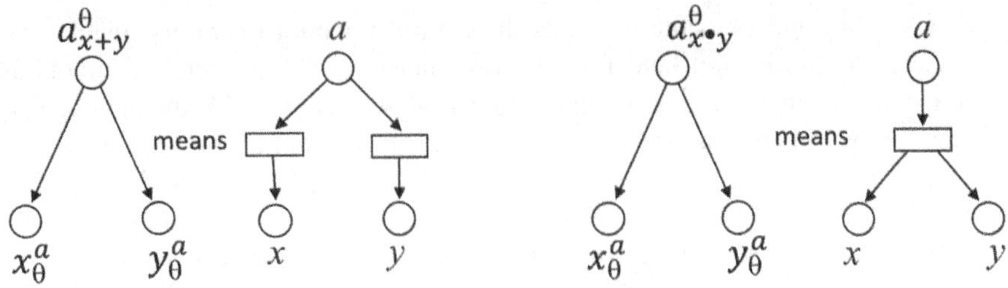


**Fig. 1.1** C-e structures and behaviourally equivalent Petri nets when $y \neq \theta$

subscript (superscript) $\theta$ of a node symbolises lack of the node's successors (predecessors). Thus, exclusive choice between $\{a^{\theta}_{x}, x^{a}_{\theta}\}$ and the empty c-e structure, that is $\{a^{\theta}_{\theta}\}$ (the isolated node $a$), should entail sending token from $a$ to $x$, but not only removing it from $a$. General discussions on interpretation of various kinds of choice between actions is in [1, 7, 10, 15, 16].

6. A useful remark for calculations with formal polynomials concerns their transforming into the canonical form, that is a repeated sum (+) of products ($\bullet$) of node symbols: each polynomial $K \in \boldsymbol{F}[\mathbb{X}]$ may be transformed by equations in Definition 1.1, into this form. Evidence–by structural induction: assume that $L$ and $M$ are in canonical form, if $K = L + M$ or $K = L \bullet M$ then using equations (+)–(+$\bullet$) brings $K$ to the sum of products.
7. Since functions $C$ and $E$ are total, each c-e structure comprises all nodes from $\mathbb{X}$, also the isolated ones–those from outside of its carrier, i.e. with $C(x) = E(x) = \theta$, invisible in the graphic presentation. The carrier only is visible. The isolated nodes make the distributivity law (+$\bullet$) conditional. A representation of a c-e structure $U = \langle C, E \rangle$ as a set of annotated nodes is $\{x^{C(x)}_{E(x)} : x \in car(U)\}$. Pictorially, $U$ is a directed graph with set of vertices $car(U)$ labelled with objects of the form $x^{C(x)}_{E(x)}$ ($x \in car(U)$); there is an edge from $x$ to $y$ iff $y$ occurs in the polynomial $E(x)$. Notice that in this representation, edges (arrows), although useful for the graphical look-out, are redundant: interconnection of vertices follows from the structure of polynomials $C(x)$, $E(x)$. In some examples we shall omit $C(x)$ or $E(x)$ if they are $\theta$.

Figure 1.3a and b depict two graphical presentations of the same c-e structure with carrier $\{a, b, c, d, e, f, g, h, x\}$; in (a) the encircled nodes comprise groups making products in formal polynomials in (b), where the sums of the products create families of the groups. This c-e structure is the set

$$\{a^{\theta}_{x}, b^{\theta}_{x}, c^{\theta}_{x}, d^{\theta}_{x}, x^{a+b\bullet c+c\bullet d}_{e\bullet f+f\bullet g+h}, e^{x}_{\theta}, f^{x}_{\theta}, g^{x}_{\theta}, h^{x}_{\theta}\}.$$

**Definition 1.3** (*addition and multiplication; monomial c-e structure*) For c-e structures $U = \langle C_U, E_U \rangle$, $V = \langle C_V, E_V \rangle$ define:

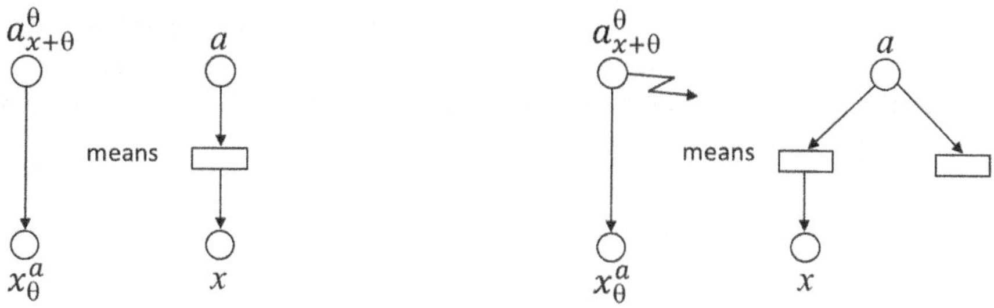


**Fig. 1.2** Adding neutral $\theta$–interpretation in terms of Petri nets

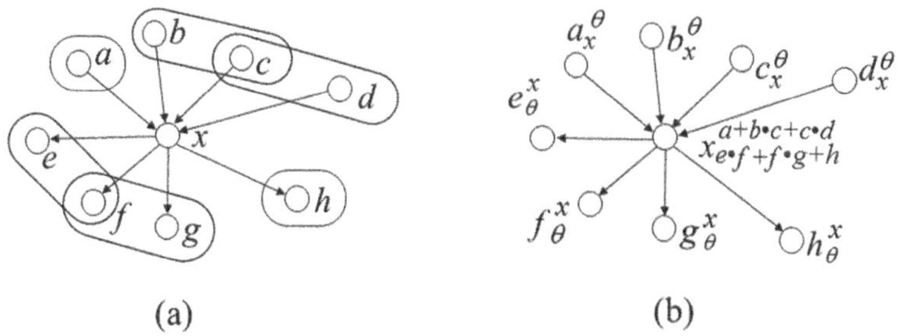


**Fig. 1.3** **a** Predecessors and successors of the node $x$, grouped into families: $\{\{a\}, \{b, c\}, \{c, d\}\}$ and $\{\{e, f\}, \{f, g\}, \{h\}\}$. **b** Notation by means of formal polynomials

$$
\begin{aligned}
U + V = \ & \langle C_{U+V}, E_{U+V}\rangle = \langle C_U + C_V, E_U + E_V\rangle,\\
& \text{where } (C_U + C_V)(x) = C_U(x) + C_V(x), \text{ and similarly for} E;\\
U \bullet V = \ & \langle C_{U\bullet V}, E_{U\bullet V}\rangle = \langle C_U \bullet C_V, E_U \bullet E_V\rangle,\\
& \text{where } (C_U \bullet C_V)(x) = C_U(x) \bullet C_V(x), \text{ and similarly for} E.
\end{aligned}
$$

The same symbols "+" and "•" are used for operations on c-e structures and formal polynomials, so if $A \subseteq \boldsymbol{CE}$ then $\Sigma A$ denotes the sum $(+)$[2] of all c-e structures in $A$. $U$ is a *monomial* c-e structure iff each polynomial $C_U(x)$ and $E_U(x)$ is a monomial, i.e. does not comprise non-reducible (relative to equations in Definition 1.1) operation "+". Obviously $U + V$ and $U \bullet V$ are correct, since $x$ occurs in $C_{U+V}(y)$ iff $y$ occurs in $E_{U+V}(x)$ and the same for $U \bullet V$. □

Apart from the representation of c-e structures as a set $\{x^{C(x)}_{E(x)} : x \in car(U)\}$, their linear notation is used as the so-called "*arrow expressions*": $\{x^{\theta}_{y}, y^{x}_{\theta}\}$ is an arrow, denoted as $x \to y$ and, consequently, $\{x^{\theta}_{y}, y^{x}_{\theta}\} \bullet \{y^{\theta}_{z}, z^{y}_{\theta}\} \bullet \{z^{\theta}_{u}, u^{z}_{\theta}\} \ldots = \{x^{\theta}_{y}, y^{x}_{z}, z^{y}_{u}, u^{z}_{\ldots} \ldots\}$ is a chain, denoted as $x \to y \to z \to u \ldots$. Bidirectional arrow

[2] Although the sum $(+)$ of any number of c-e structures turns out to be their least upper bound relative to the partial order introduced in Definition 1.4, their product $(\bullet)$ has nothing to do with the greatest lower bound. So, the symbol $\Pi A$ will be used for the greatest lower bound of the set $A$, not for repeated multiplication $(\bullet)$ of c-e structures. Recall point 4 of remarks commenting Definitions 1.1 and 1.2. The lattice operations $\Sigma A$ and $\Pi A$, thus for infinite sets $A$ too, will be defined later.

$x \leftrightarrow y$ denotes $x \to y \to x$ (equivalent to $y \to x \to y$), that is, the close cycle $\{x_y^y, y_x^x\}$. Chains and arrows in particular, may be combined into "arrow expressions" representing some c-e structures. For instance c-e structure $\{a_{x+y}^{\theta}, b_{x\bullet y}^{\theta}, x_{\theta}^{a\bullet b}, y_{\theta}^{a\bullet b}\}$ may be written as $(a \to x + a \to y) \bullet (b \to x) \bullet (b \to y)$.

The set $\boldsymbol{CE}$ with addition, multiplication and a distinguished c-e structure $\langle C_\theta, E_\theta\rangle$ such that $C_\theta(x) = E_\theta(x) = \theta$ for all $x \in \mathbb{X}$, makes an algebraic system similar to that in Definition 1.1, the quasi-semiring of c-e structures. Denoting $\langle C_\theta, E_\theta\rangle$ also by $\theta$ (and complying with a role $\theta$ plays in a concrete context),[3] we have:

**Proposition 1.1** *The system $\langle \boldsymbol{CE}[\mathbb{X}], +, \bullet, \theta\rangle$ obeys the following equations for all $U, V, W \in \boldsymbol{CE}[\mathbb{X}]$, $x, y \in \mathbb{X}$:*

| | | | |
|---|---|---|---|
| (+) | $\theta + U = U + \theta = U$ | (•) | $\theta \bullet U = U \bullet \theta = U$ |
| (++) | $U + U = U$ | (••) | $(x \to y) \bullet (x \to y) = x \to y$ |
| (+++) | $U + V = V + U$ | (•••) | $U \bullet V = V \bullet U$ |
| (++++) | $U + (V + W) = (U + V) + W$ | (••••) | $U \bullet (V \bullet W) = (U \bullet V) \bullet W$ |
| (+•) | *If* $C_V(x) \neq \theta \Leftrightarrow C_W(x) \neq \theta$ *and* $E_V(x) \neq \theta \Leftrightarrow E_W(x) \neq \theta$, *then* $U \bullet (V + W) = U \bullet V + U \bullet W$. | | |

□

**Definition 1.4** (*partial order $\leq$; substructure; set $\boldsymbol{SUB}[V]$; firing component; set $\boldsymbol{FC}$; pre-set and post-set*) For $U, V \in \boldsymbol{CE}$ let $U \leq V \Leftrightarrow V = U + V$; $\leq$ is a partial order in $\boldsymbol{CE}$. If $U \leq V$ then $U$ is a *substructure* of $V$; $\boldsymbol{SUB}[V] = \{U : U \leq V\}$ is the set of all substructures of $V$. For $A \subseteq \boldsymbol{CE}$: $V \in A$ is *minimal* (w.r.t. $\leq$) in $A$ iff $\forall W \in A : (W \leq V \Rightarrow W = V)$.

A minimal in $\boldsymbol{CE}\backslash\{\theta\}$ c-e structure $Q = \langle C_Q, E_Q\rangle$ is a *firing component* iff $Q$ is a monomial c-e structure and $C_Q(x) = \theta \Leftrightarrow E_Q(x) \neq \theta$ for any $x \in car(Q)$. The set of all firing components is denoted by $\boldsymbol{FC}$, thus the set of all firing components of $U \in \boldsymbol{CE}$ is $\boldsymbol{FC}[U] = \boldsymbol{SUB}[U] \cap \boldsymbol{FC}$. Let for $Q \in \boldsymbol{FC}$ and $G \subseteq \boldsymbol{FC}$:

$$
\begin{aligned}
{}^\bullet Q &= \{x \in car(Q) : C_Q(x) = \theta\} && (\textit{pre-set} \text{ or causes of } Q)\\
Q^\bullet &= \{x \in car(Q) : E_Q(x) = \theta\} && (\textit{post-set} \text{ or effects of } Q)\\
{}^\bullet Q^\bullet &= {}^\bullet Q \cup Q^\bullet &&\\
{}^\bullet G &= \bigcup_{Q \in G} {}^\bullet Q && (\textit{pre-set} \text{ or causes of } G)\\
G^\bullet &= \bigcup_{Q \in G} Q^\bullet && (\textit{post-set} \text{ or effects of } G)\\
{}^\bullet G^\bullet &= {}^\bullet G \cup G^\bullet &&
\end{aligned}
$$

□

Thus, the firing component is a connected graph, due to the required minimality. Elements of the pre-set are its *causes* and elements of the post-set are its *effects*.

**Definition 1.5** (*volley–pairwise detached firing components; family $\boldsymbol{FCV}$*) Firing components $Q$ and $P$ are *detached* if and only if ${}^\bullet Q^\bullet \cap {}^\bullet P^\bullet = \varnothing$. Any set $G \subseteq \boldsymbol{FC}$

[3] Thus, every node of the c-e structure $\theta$ is isolated: not linked with any other.

of pairwise detached firing components is called their **volley**.[4] The family of volleys is denoted by $\boldsymbol{FCV}$. So, if $G \subseteq \boldsymbol{FC}[U]$ then $\boldsymbol{FCV}[U] \subseteq 2^{\mathbf{FC}[U]}$ for a c-e structure $U$, denotes a collection of volleys in $U$. The intention is that firing components in a volley should act ("fire") simultaneously. This notion will be used in definition of parallel semantics of c-e structures. □

**Definition 1.6** (*state of elementary c-e structures*) A *state* is a subset of the set of nodes: $s \subseteq \mathbb{X}$. The set of all states: $\mathbb{S} = 2^{\mathbb{X}}$. A node $x$ is *active* in the state $s$ iff $x \in s$ and *passive* otherwise. As in case of Petri nets, we say "$x$ holds a token" when $x$ is active. Obviously, the state might be defined equivalently as a two-valued function $s : \mathbb{X} \to \{\mathbf{0}, \mathbf{1}\}$. □

**Definition 1.7** (*sequential and parallel semantics of elementary c-e structures*)

**Sequential** For $Q \in \boldsymbol{FC}[U]$, $s \in \mathbb{S}$, let $[\![Q]\!] \subseteq \mathbb{S} \times \mathbb{S}$ be a binary relation defined as:

$$(s, t) \in [\![Q]\!] \quad \text{iff} \quad {}^{\bullet}Q \subseteq s \wedge Q^{\bullet} \cap s = \varnothing \wedge t = (s \setminus {}^{\bullet}Q) \cup Q^{\bullet}$$

($Q$ transforms state $s$ into $t$). *Semantics* $[\![U]\!]$ of $U \in \boldsymbol{CE}$ is:

$$[\![U]\!] = \bigcup_{Q \in FC[U]} [\![Q]\!].$$

Its reflexive and transitive closure is denoted by $[\![U]\!]^*$, so, $(s, t) \in [\![U]\!]^*$ if and only if $s = t$ or there exists a sequence of states $s_0, s_1, \ldots, s_n$ with $s = s_0$, $t = s_n$ and $(s_j, s_{j+1}) \in [\![U]\!]$ for $j = 0, 1, \ldots, n-1$. State $t$ is *reachable* from $s$ in semantics $[\![U]\!]$ and the sequence $s_0, s_1, \ldots, s_n$ is a *computation* performed by $U$.

**Parallel** For a volley $G \in \boldsymbol{FCV}[U]$, $G \neq \varnothing$, relations $[\![G]\!]$ and $[\![U]\!]_{par}$ are defined in the same way as $[\![Q]\!]$ and $[\![U]\!]$ in the sequential case but with $Q$ replaced with $G$ and $\boldsymbol{FC}[U]$ replaced with $\boldsymbol{FCV}[U]$. Closure $[\![U]\!]^*_{par}$, reachability and computation are defined as in the sequential case. □

Note that $[\![U]\!] = \varnothing$ iff $\boldsymbol{FC}[U] = \varnothing$, which is a structural deadlock, and that the firing component is a two level bipartite connected graph with $n > 1$ causes and $m > 1$ effects. Firing components with the same pre-set and the same post-set have identical semantics, thus behave identically: if ${}^{\bullet}Q = {}^{\bullet}P$ and $Q^{\bullet} = P^{\bullet}$ then $[\![Q]\!] = [\![P]\!]$. $Q$ and $P$ are behaviourally equivalent. For instance, there are 5 equivalent firing components with $n = 2$ and $m = 2$, these are the following: $\{a^{\theta}_{x \bullet y}, b^{\theta}_{y}, x^{a}_{\theta}, y^{a \bullet b}_{\theta}\}$, $\{a^{\theta}_{x}, b^{\theta}_{x \bullet y}, x^{a \bullet b}_{\theta}, y^{b}_{\theta}\}$, $\{a^{\theta}_{y}, b^{\theta}_{x \bullet y}, x^{b}_{\theta}, y^{a \bullet b}_{\theta}\}$, $\{a^{\theta}_{x \bullet y}, b^{\theta}_{x}, x^{a \bullet b}_{\theta}, y^{a}_{\theta}\}$, $\{a^{\theta}_{x \bullet y}, b^{\theta}_{x \bullet y}, x^{a \bullet b}_{\theta}, y^{a \bullet b}_{\theta}\}$. The latter is called *closely connected*. Other properties are listed below. Behaviour of elementary c-e structures may be imagined as a token game: if each node in a firing component's pre-set holds a token and none in its post-set does, then remove tokens from the pre-set and put them in the post-set. This is illustrated in Fig. 1.4 where evolution of the state of this c-e structure $U$ is exampled

[4] In [6] instead of the word "volley", the name "salvo" has been used.

by distribution of tokens amids nodes in successive stages of its work. The set of firing components of this c-e structure is

$$\boldsymbol{FC}[U] = \{\{a_b, b^a\}, \{b_{c\bullet d}, c^b, d^b\}, \{b_e, e^b\}, \{c_d, d^c\}, \{c_e, e^c\}, \{e_a, a^e\}, \{e_d, d^e\}\}$$

The sum $\sum \boldsymbol{FC}[U]$ of all these firing components is $U$. But this is not the general property, as will be seen later. Note that firing components are counterparts of transitions in Petri nets.

**Example 1.1 (node neighbours)**

In this example the concept of neighbourhood, crucial in the next chapters, is illustrated by rules of admissible motions of chess pieces on the board. The chessboard is represented by a c-e structure in the graphic form, whose nodes labelled by $a[x, y]$ correspond to the square fields of the board. In Fig. 1.5 the motion of the chess knight progresses along the arrows connecting nodes. Two nodes are linked with bidirectional arrow if a move between them by the knight is allowed. They are *neighbours* of each other. For $3 \leq x \leq 6$ and $3 \leq y \leq 6$ the nodes $a[x, y]$ are linked to eight neighbours: $a[x-2, y-1], a[x-2, y+1], a[x-1, y+2], a[x+1, y+2]$, $a[x+2, y+11]$, $a[x+2, y-1]$, $a[x+1, y-2]$, $a[x-1, y-2]$ which are causes and effects of $a[x, y]$. They make a *neighbourhood* of $a[x, y]$. Obviously, nodes nearer to the borders have fewer neighbours. How many?–it is seen at once in Fig. 1.5. So, this c-e structure $U = \{a[x, y]^{\text{sum of neighbours}}_{\text{sum of neighbours}} : a[x, y] \in car(U)\}$ with carrier $car(U) = \{a[x, y] : 1 \leq x \leq 8 \wedge 1 \leq y \leq 8\}$ has the set of firing components $\boldsymbol{FC}[U]$ being the collection of ordered pairs of neighbouring nodes, therefore linked with an arrow. Here too $\sum \boldsymbol{FC}[U] = U$, which is directly evident, since no multiplication "•" occurs in $U$: any chess figure can go to one square only at a time. It is seen that any node is reachable from any other.

In Fig. 1.6 the motion of the chess king progresses along the arrows connecting nodes. The carrier of this c-e structure is the same, the neighbours of nodes and the firing components are readily perceived from the picture.

Easy to verify properties of elementary c-e structures, inferred from above definitions and proved in [5] are in the following Propositions.

**Proposition 1.2** (structural properties) *For $U, V, W \in \boldsymbol{CE}$:*

1. $U \bullet (V + W) \leq U \bullet V + U \bullet W$.
2. *If* $U \bullet (V + W) = U \bullet V + U \bullet W$, *then* $V \leq W \Rightarrow U \bullet V \leq U \bullet W$.
3. $(U \bullet V)^n = U^n \bullet V^n$ *(where* $U^0 = \theta$, $U^{n+1} = U \bullet U^n$*) for* $n = 0, 1, 2, \ldots$.
4. *If* $C_U(x) \neq \theta \Leftrightarrow C_V(x) \neq \theta$ *and* $E_U(x) \neq \theta \Leftrightarrow E_V(x) \neq \theta$ *for any* $x$, *then* $(U + V)^n = \sum_{i=0}^{n} U^{n-i} \bullet V^i$.
5. *If* $car(U) \cap car(V) = \emptyset$ *and* $V \leq W$, *then* $V \leq U \bullet W$.
6. $U \leq V \Leftrightarrow \boldsymbol{SUB}[U] \subseteq \boldsymbol{SUB}[V]$.
7. $\boldsymbol{SUB}[U] \cup \boldsymbol{SUB}[V] \subseteq \boldsymbol{SUB}[U + V]$.
8. $\boldsymbol{SUB}[U] \cup \boldsymbol{SUB}[V] = \boldsymbol{SUB}[U + V] \Leftrightarrow (U \leq V \vee V \leq U)$.
9. $U \leq V \Rightarrow \boldsymbol{FC}[U] \subseteq \boldsymbol{FC}[V]$, *but converse implication not always holds.*

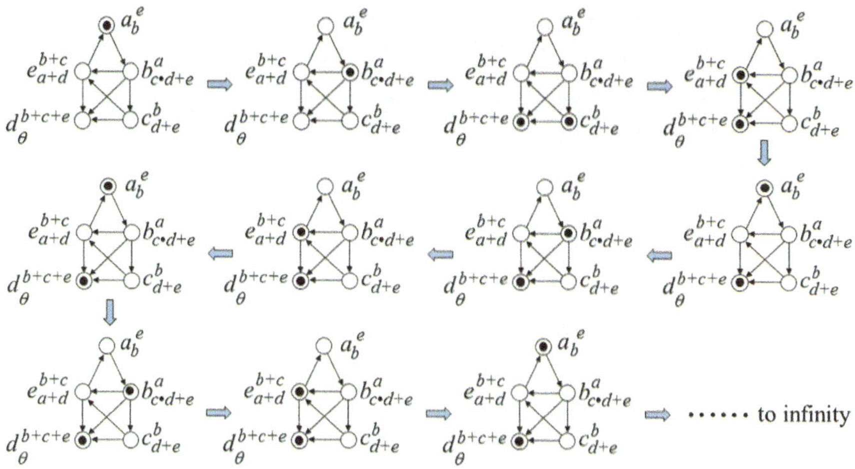


**Fig. 1.4** Possible successive transformations of states during activity of c-e structure $U$ in semantics $[\![U]\!]$. Its linear notation as an "arrow expression"is the following: $(a \to b) + (b \to c) \bullet (b \to d) + (b \to e) + (c \to d) + (c \to e) + (e \to a) + (e \to d)$

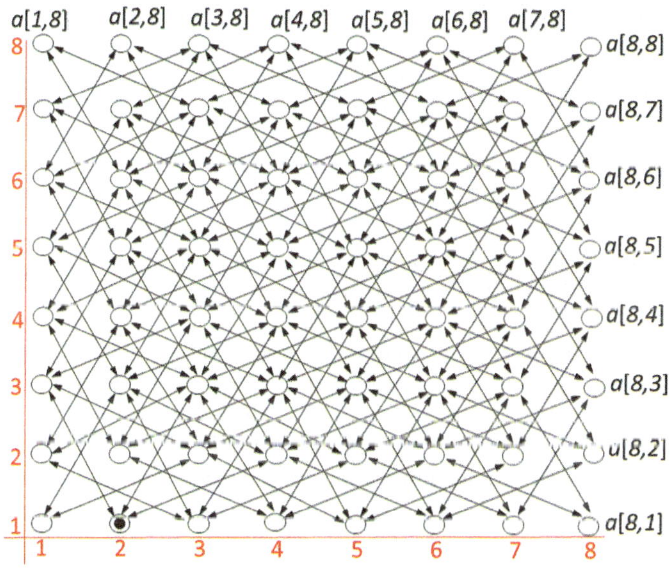


**Fig. 1.5** C-e structure showing paths of motion of the knight on the board

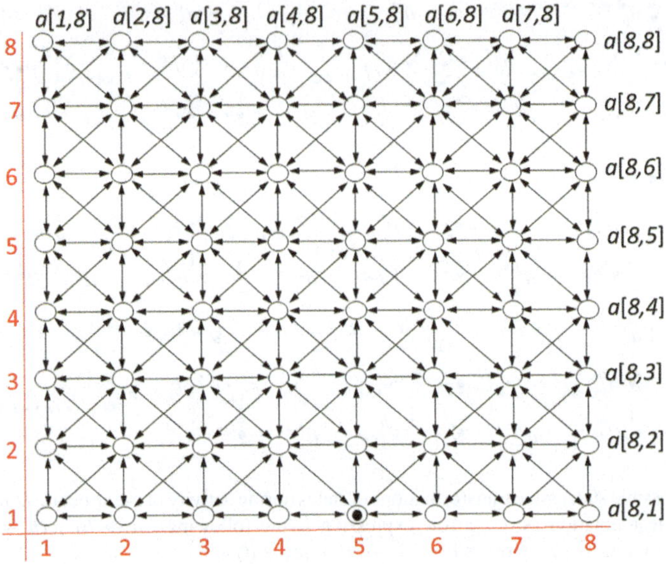


**Fig. 1.6** C-e structure showing paths of motion of the king on the board

*10.* $\boldsymbol{FC}[U] \cup \boldsymbol{FC}[V] \subseteq \boldsymbol{FC}[U+V]$, *but inclusion not always can be replaced with equality.*

*11.* $\boldsymbol{SUB}[U] \cup \boldsymbol{SUB}[V] = \boldsymbol{SUB}[U+V] \Rightarrow \boldsymbol{FC}[U] \cup \boldsymbol{FC}[V] = \boldsymbol{FC}[U+V]$, *but converse implication not always holds.*

*12.* *Each firing component* $Q \in \boldsymbol{FC}$ *is a directed and connected bipartite graph with* $k$ *arrows beginning in the pre-set* ${}^\bullet Q$ *and ending in the post-set* $Q^\bullet$, *where* $k$ *satisfies* $|{}^\bullet Q| + |Q^\bullet| - 1 \leq k \leq |{}^\bullet Q| \cdot |Q^\bullet|$. □

Point 10 in Proposition 1.2 needs some comment. It states that new firing components may appear when summing up c-e structures. For instance, let $U = \{a_{x+y}, b_{x\bullet y}, x^{a\bullet b}, y^{a\bullet b}\}$, $V = \{a_{x\bullet y}, x^a, y^a\}$, thus $\boldsymbol{FC}[U] = \varnothing$, $\boldsymbol{FC}[V] = \{V\}$, $\boldsymbol{FC}[U+V] = \{\{a_x, x^a\}, \{a_y, y^a\}, V, \{a_{x\bullet y}, b_{x\bullet y}, x^{a\bullet b}, y^{a\bullet b}\}\}$, thus $\boldsymbol{FC}[U] \cup \boldsymbol{FC}[V] \neq \boldsymbol{FC}[U+V]$. Eventuality of appearing new firing components when assembling c-e structures from smaller parts, reflects a general observation: compound systems may sometimes reveal behaviours absent in their parts, though retain behaviour of the parts. What may also be commented is that creation of new firing components in effect of summation seems rather exceptional and were not experienced with realistic examples of systems design by means of c-e structures. In this context one may notice that the formal means of combining system models, though facilitate or enable to derive behavioural properties of large structures out of smaller (descriptive compositionality),[5] may neglect some new behaviours. This "act of creation" appearing in real processes, like the synergetic effect in economy (where common activity of a group of firms or agents yields greater total effect than separate activities), or in chemistry, may hinder formal analysis of behaviour, thus may oppose the principle of compositionality. As a remedy, identification of some easy-to-verify conditions ensuring absence of new firing components created in summation has been proposed: a conditional compositionality. This concerns the standard c-e structures but not their cellular version, where is only addition.

**Proposition 1.3** (semantical properties) *For any firing components $Q, P \in \boldsymbol{FC}$, non-empty sets $G, H \subseteq \boldsymbol{FC}$ of pairwise detached firing components, and any c-e structures $U, V \in \boldsymbol{CE}$:*

1. $[\![Q]\!] \neq \varnothing$, $[\![G]\!]_{par} \neq \varnothing$.
2. $[\![Q]\!] \neq [\![P]\!] \Leftrightarrow [\![Q]\!] \cap [\![P]\!] = \varnothing$, $[\![G]\!]_{par} \neq [\![H]\!]_{par} \Leftrightarrow [\![G]\!]_{par} \cap [\![H]\!]_{par} = \varnothing$.
3. $U \leq V \Rightarrow \boldsymbol{FC}[U] \subseteq \boldsymbol{FC}[V] \Rightarrow [\![U]\!] \subseteq [\![V]\!] \Rightarrow [\![U]\!]^* \subseteq [\![V]\!]^*$, *but none of the implications "$\Rightarrow$" may be replaced with equivalence "$\Leftrightarrow$".*
4. $\boldsymbol{FC}[U] \subseteq \boldsymbol{FC}[V] \Rightarrow [\![U]\!]_{par} \subseteq [\![V]\!]_{par} \Rightarrow [\![U]\!]^*_{par} \subseteq [\![V]\!]^*_{par}$, *but none of the implications "$\Rightarrow$" may be replaced with equivalence "$\Leftrightarrow$".*
5. $[\![U]\!] \subseteq [\![U]\!]_{par}$, *but the reverse inclusion not always holds.*
6. $[\![U]\!]^* = [\![U]\!]^*_{par}$.
7. $[\![U]\!] \cup [\![V]\!] \subseteq [\![U + V]\!]$, *but the reverse inclusion not always holds.*
8. $[\![U]\!] \cup [\![V]\!] \subseteq [\![U]\!]^* \cup [\![V]\!]^* \subseteq ([\![U]\!] \cup [\![V]\!])^* \subseteq [\![U + V]\!]^*$, *but none of the inclusions "$\subseteq$" may be replaced with equality "$=$".*
9. $\boldsymbol{FC}[U] \cup \boldsymbol{FC}[V] = \boldsymbol{FC}[U + V] \Rightarrow [\![U]\!] \cup [\![V]\!] = [\![U + V]\!]$, *but not conversely.*
10. $\boldsymbol{FC}[U] \cup \boldsymbol{FC}[V] = \boldsymbol{FC}[U + V]$ *and* $[\![U]\!]^* \cup [\![V]\!]^* = [\![U + V]\!]^*$ *are unrelated by implication.*
11. $\boldsymbol{FC}[U] \cup \boldsymbol{FC}[V] = \boldsymbol{FC}[U + V]$ *and* $[\![U]\!]_{par} \cup [\![V]\!]_{par} = [\![U + V]\!]_{par}$ *are unrelated by implication.*
12. $\boldsymbol{FC}[U] \cup \boldsymbol{FC}[V] = \boldsymbol{FC}[U + V]$ *and* $[\![U]\!]^*_{par} \cup [\![V]\!]^*_{par} = [\![U + V]\!]^*_{par}$ *are unrelated by implication.* □

Point 7 in Proposition 1.3 states that sum of c-e structures, apart from behaviours present in its constituents, may exhibit behaviours absent in them. For instance, if $U = \{a_{x+y}, b_{x\bullet y}, x^{a\bullet b}, y^{a\bullet b}\}$, $V = \{a_{x\bullet y}, x^a, y^a\}$, then $[\![U]\!] = \varnothing$, $[\![V]\!] = [\![\{a_{x\bullet y}, x^a, y^a\}]\!]$, $[\![U + V]\!] = [\![\{a_{x\bullet y}, x^a, y^a\}]\!] \cup [\![\{a_x, x^a\}]\!] \cup [\![\{a_y, y^a\}]\!] \cup [\![\{a_{x\bullet y}, b_{x\bullet y}, x^{a\bullet b}, y^{a\bullet b}\}]\!]$. Thus, $[\![U]\!] \cup [\![V]\!] \neq [\![U + V]\!]$. This means that, the semantics, in general, is not fully compositional, but point 9 gives an obvious sufficient condition for being such.

So far the summation of finite number of c-e structures was considered but this may be extended to infinity. This is especially advisable because their cellular version will be often constructed as infinite sums of finite small c-e structures called "atomic". The least upper bound and greatest lower bound of a set of c-e structures are defined by means of the partial order $\leq$ (Definition 1.4). Although the greatest lower bound exists for any set of c-e structures, this is not so with the least upper bound.

[5] Note that semantics of c-e structures has been defined by means of a structural entity–the firing component, not by extraneous objects from a certain model.

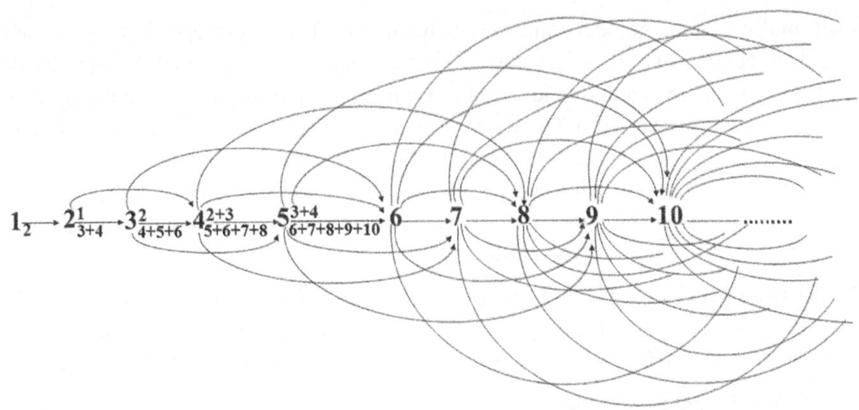


**Fig. 1.7** Infinite c-e structure with $car(U) = \mathbb{X} =$ *set of positive integers*. The cause and effect polynomials annotating nodes are finite but of unbounded length

**Definition 1.8** (*least upper and greatest lower bound*) For a set of c-e structures $A \subseteq \boldsymbol{CE}$, the least upper bound of $A$, denoted by $lub(A)$, is the least (with respect to $\leq$) c-e structure $V$ with $U \leq V$ for each $U \in A$. Since for finite $A$, $lub(A)$ amounts to the sum $\sum A$, let us retain this denotation also for infinite $A$. The greatest lower bound of $A$, denoted by $glb(A)$ or $\prod A$, is the greatest c-e structure $V$ with $V \leq U$ for each $U \in A$. In addition $lub(\emptyset) = glb(\emptyset) = \theta$. □

Customary denotations for $\sum A$ are: $\sum_{j \in I} U_j = \sum\{U_j \mid j \in I\}$ ($I$ is a set of indices) or $\sum_{j=0}^{n} U_j$ when $I = \{0, 1, \ldots, n\}$ or $\sum_{j=0}^{\infty} U_j$, or $\sum\{U_1, U_2, \ldots, U_n\} = U_1 + U_2 + \ldots + U_n$, and similarly for $\prod A$. However for infinite $A$, not always exists $\sum A$. For instance, let $A = \{\{a_1, \mathbf{1}^a\}, \{a_2, \mathbf{2}^a\}, \{a_3, \mathbf{3}^a\}, \ldots\}$ (the boldface numerals are names of nodes). Obviously, $A \subseteq \boldsymbol{CE}[\mathbb{X}]$. Suppose $\sum A \in \boldsymbol{CE}[\mathbb{X}]$ and let $U_i = \{a_\mathbf{i}, \mathbf{i}^a\}$ $(i = 1, 2, 3, \ldots, \mathbf{i} = \mathbf{1}, \mathbf{2}, \mathbf{3}, \ldots)$ and $S_i = U_1 + U_2 + \ldots + U_i$. Hence, the subscript polynomial of $a$ in $S_{i+1}$ is longer than that in $S_i$. Since each $S_i \leq \sum A$ then the subscript polynomial of $a$ in $\sum A$ would be of infinite length: $\mathbf{1} + \mathbf{2} + \mathbf{3} + \ldots$ thus, $\sum A \notin \boldsymbol{CE}[\mathbb{X}]$ – a contradiction. There exist, however, c-e structures with unbounded length of polynomials. An example is:

$$U = \{\mathbf{1_2}, \mathbf{2^1_{3+4}}, \mathbf{3^2_{4+5+6}}, \mathbf{4^{3+2}_{5+6+7+8}}, \mathbf{5^{4+3}_{6+7+8+9+10}}, \mathbf{6^{5+4+3}_{7+8+9+10+11+12}}, \ldots\}$$

depicted in Fig. 1.7.

Some properties of the least upper bound and greatest lower bound are collected in Proposition 1.4. One should bear in mind that the operators $\sum$ and $\prod$ are defined by means of the ordering $\leq$ among c-e structures, but merely $\sum$ generalizes addition of c-e structures to infinite number of addends. The operator $\prod$ does not concern multiplication ($\bullet$) of infinite number of factors. The customary mathematical operators $\bigcup$ and $\bigcap$ in Proposition 1.4, sometimes called "join" and "meet", generalize union and intersection of sets to infinite number of constituents.

**Proposition 1.4** (properties of *lub* and *glb*)

1. *For each $A \subseteq$ **CE** $\prod A$ exists and $\prod A \in$ **CE**, thus $x$ occurs in $C_{\prod A}(y)$ iff $y$ occurs in $E_{\prod A}(x)$ (Definition 1.2). $\sum A \in$ **CE** if $\sum A$ exists.*
2. *For each $A \subseteq$ **CE**, if $\sum A$ exists then $A \subseteq \boldsymbol{SUB}[\sum A]$.*
3. *For each $U \in$ **CE** there exists $\sum \boldsymbol{SUB}[U]$ and $\sum \boldsymbol{SUB}[U] = U$.*
4. $\bigcup_{j \in I} \boldsymbol{SUB}[U_j] \subseteq \boldsymbol{SUB}[\sum_{j \in I} U_j]$.
5. $\sum_{j \in I} U_j = \sum \bigcup_{j \in I} \boldsymbol{SUB}[U_j]$ *provided that least upper bounds involved exist.*
6. $\prod_{j \in I} U_j = \sum \bigcap_{j \in I} \boldsymbol{SUB}[U_j]$ *provided that least upper bound involved exists.*
7. *If $A \subseteq \boldsymbol{SUB}[U]$ then there exists $\sum A$ in $\boldsymbol{SUB}[U]$.*
8. *For any sets $A, B \subseteq$ **CE**: $\sum(A \cup B) = \sum A + \sum B$ provided that all the least upper bounds involved exist.*
9. *For any sets $A, B \subseteq$ **CE**: $A \subseteq B \Rightarrow \sum A \leq \sum B$ provided that $\sum A$ and $\sum B$ exist.*
10. $\forall j \in I : (U_j \leq V_j \Rightarrow \sum_{j \in I} U_j \leq \sum_{j \in I} V_j$ *if* $\sum_{j \in I} U_j$ *and* $\sum_{j \in I} V_j$ *exist.*
11. $\sum$ ***CE*** *does not exist.*
12. $\prod \boldsymbol{CE} = \theta$.
13. *For each $A \subseteq$ **FC**, if $\sum A$ exists then $A \subseteq \boldsymbol{FC}[\sum A]$.*
14. *For each $U \in$ **CE** there exists $\sum \boldsymbol{FC}[U]$ and $\sum \boldsymbol{FC}[U] \leq U$. Sometimes $\sum \boldsymbol{FC}[U] \neq U$.*
15. $\bigcup_{j \in I} \boldsymbol{FC}[U_j] \subseteq \boldsymbol{FC}[\sum_{j \in I} U_j]$ *(generalization of 10 in Proposition 1.2).*
16. $\bigcup_{j \in I} \boldsymbol{FC}[U_j] = \boldsymbol{FC}[\sum_{j \in I} U_j] \Rightarrow \bigcup_{j \in I} [\![U_j]\!] = [\![\sum_{j \in I} U_j]\!]$ *(generalization of 9 in Proposition 1.3).*
17. *If $\sum \boldsymbol{FC}[U] = U$ then $\boldsymbol{FC}[U] \subseteq \boldsymbol{FC}[V] \Rightarrow U \leq V$, for each $V \in$ **CE**.*
18. $[\![\sum \boldsymbol{FC}[U]]\!] = [\![U]\!]$. □

The concepts and properties collected so far imply that the set of all c-e structures partially ordered as in Definition 1.4 make a lattice, that is for any $U, V \in$ **CE**, there exist $\sum\{U, V\}$ and $\prod\{U, V\}$ in **CE**. Obviously, $\sum\{U, V\} = U + V$ and a special case of points 5 and 6 in Proposition 1.4 are equations[6]:

$$\sum\{U, V\} = \sum(\boldsymbol{SUB}[U] \cup \boldsymbol{SUB}[V])$$
$$\prod\{U, V\} = \sum(\boldsymbol{SUB}[U] \cap \boldsymbol{SUB}[V])$$

Some properties of the lattice of c-e structures can be stated in the following proposition:

**Proposition 1.5** (non-distributive and incomplete lattice) *The lattice $\langle$**CE**$, \sum, \prod\rangle$ is non-distributive, that is not always $\prod\{U, \sum\{V, W\}\} = \sum\{\prod\{U, V\}, \prod\{U, W\}\}$ holds. Its least element is $\theta$, the greatest does not exist; this is not complete lattice.*[7] □

---

[6] Remember that $\prod\{U, V\}$ does not mean $U \bullet V$.

[7] A lattice is distributive if $\prod\{U, \sum\{V, W\}\} = \sum\{\prod\{U, V\}, \prod\{U, W\}\}$ and is complete if $\sum A$ and $\prod A$ exist and belong to it, for any of its non-empty subset $A$. Remember that $\theta$ is the short notation of the neutral c-e structure $\langle C_\theta, E_\theta\rangle$, as used in Proposition 1.1.

Sometimes it is convenient to write $\sum\{U, V\}$ and $\prod\{U, V\}$ in the infix form: $U + V$ and $U \circ V$, as the associative and commutative two-argument operations. In terms of these operations, some properties concerning lattices of c-e structures are expressed[8]:

**Proposition 1.6** (properties of "$\circ$" as compared to "$\bullet$")

1. *$U \circ (V + W) \geq U \circ V + U \circ W$ (compare point 1 in Proposition 1.2).*
2. $U \circ (V + W) = U \circ V + U \circ W \Leftrightarrow U + (V \circ W) = (U + V) \circ (U + W)$.
3. *If $\boldsymbol{SUB}[V] \cup \boldsymbol{SUB}[W] = \boldsymbol{SUB}[V + W]$ then $U \circ (V + W) = U \circ V + U \circ W$.*
4. *$U \leq V \Leftrightarrow U = U \circ V$ (remind that by definition $U \leq V \Leftrightarrow V = U + V$).*
5. *$\boldsymbol{FC}[U] \cap \boldsymbol{FC}[V] = \boldsymbol{FC}[U \circ V]$ (compare point 10 in Proposition 1.2).*
6. *For each $U \in \boldsymbol{CE}$, $\langle \boldsymbol{SUB}[U], +, \circ \rangle$ is a non-distributive, complete lattice with the least and greatest elements $\theta$ and $U$ respectively.* □

**Remark**
If $U \circ (V + W) = U \circ V + U \circ W$ holds for all c-e structures in a lattice, then this lattice is distributive. The duality expressed in point 2 of Proposition 1.6 is a general property of lattices. Point 3 gives a sufficient condition for lattice to be distributive. Point 4 asserts representation of any lattice of c-e structures with operations $+$, $\circ$ by set of subsets of its carrier with union and intersection.

**Proposition 1.7** (as compared with Proposition 1.1) *For all $U, V, W \in \boldsymbol{CE}[\mathbb{X}]$:*

($\circ$) $\theta \circ U = U \circ \theta = \theta$
($\circ\circ$) $U \circ U = U$
($\circ\circ\circ$) $U \circ V = V \circ U$
($\circ\circ\circ\circ$) $U \circ (V \circ W) = (U \circ V) \circ W$
($+\circ$) $\boldsymbol{SUB}[V] \cup \boldsymbol{SUB}[W] = \boldsymbol{SUB}[V + W] \Rightarrow U \circ (V + W) = U \circ V + U \circ W$

In the cellular c-e structures, the main subject-matter of the next chapters, does not appear the multiplication "$\bullet$" postulated in Definition 1.1 for formal polynomials and defined in Definition 1.3 for c-e structures.[9] So, the structural properties listed above in Propositions where multiplication "$\bullet$" is used explicitly, do not apply. As the constitutive structural combinators remain "+" and "$\circ$" – the least upper bound and greatest lower bound of two cellular c-e structures. Their usage in constructing cellular c-e structures is shown in the next chapters. The operator "+" in combining small structures into larger (e.g. in assembling atomic structures into complete cellular), while the "$\circ$" in cutting off fragments of flat cellular c-e structures in order to

[8] Note that multiplication "$\bullet$" composes (or synchronizes) c-e structures, whereas operation "$\circ$" removes some parts from them.

[9] Hence, any firing component in cellular c-e structures reduces to an ordered pair of nodes linked with unidirectional arrow. The behavioural role of multiplication ($\bullet$) in ordinary c-e structures, is simultaneous content change of nodes in one (or several selected) firing component. In the cellular c-e structures, the content change of all nodes (cells) simultaneously in passing from state to state, will be accomplished by a special parallel semantics.

bend them into cylindric or toroidal. It should be also noticed that some restrictive clauses appearing in properties, like in point 9 of Proposition 1.3, may be abandoned in case of cellular c-e structures, where they are fulfilled unconditionally.

### *1.1.2 B*

Essential feature of the c-e structures presented so far, is capability of combining them into complex models of realistic dynamic systems, whose behaviour, under some well defined conditions, might be inferred from their components' behaviour. So, possibility of their compositing, by means of addition (+) and multiplication (•). Then, a converse procedure is the matter of course – their decomposition. This issue turns out more complicated than the basics presented so far. Its intricacy is caused by multiplication "•", thus, in the cellular version of c-e structures, the problem is trivial: such c-e structures are decomposable. Nonetheless a few results on decomposition of standard c-e structures sheds some light on specificity of the two formalisms. Their proofs are not as straightforward as of above Propositions and all are in [5].

**Definition 1.9** (*decomposable and prime c-e structures*) A c-e structure $U$ is *decomposable* if there are c-e structures $V, W \in \boldsymbol{CE}[\mathbb{X}]$ such that $\theta \neq V \neq U \neq W \neq \theta$ and $U = V + W$ or $U = V \bullet W$. Otherwise $U$ is *non-decomposable*, called a *prime* c-e structure. $U$ is decomposable into single arrows if $U$ may be represented by an expression with arrows $x \to y$ $(x, y \in \mathbb{X})$ as operands and + and/or • as operators. □

Decomposition of c-e structures is not, in general, unique, for example $U = \{a^{\theta}_{x+y}, b^{\theta}_{x\bullet y}, x^{a\bullet b}_{\theta}, y^{a\bullet b}_{\theta}\}$ may be decomposed into single arrows either as

$$U = ((a \to x) + (a \to y)) \bullet (b \to x) \bullet (b \to y) \quad \text{or as}$$
$$U = ((a \to x) \bullet (b \to x) + (a \to y)) \bullet (b \to y)$$

The following Proposition 1.8 states that each c-e structure $U = \langle C_U, E_U \rangle$ can be transformed into the two-level c-e structure $U' = \langle C_{U'}, E_{U'} \rangle$ with $C_{U'}(x) = \theta \Leftrightarrow E_{U'}(x) \neq \theta$ for each $x \in car(U')$, which retains property of being decomposable or prime (but does not preserve behaviour, of course). $U'$ is called "two-level" because such c-e structure may be presented as a matrix of two rows, like in example:

$$\{a^{\theta}_{x+y}, b^{\theta}_{x\bullet y}, x^{a\bullet b}_{\theta}, y^{a\bullet b}_{\theta}\} = \begin{pmatrix} a^{\theta}_{x+y} & b^{\theta}_{x\bullet y} \\ & \\ x^{a\bullet b}_{\theta} & y^{a\bullet b}_{\theta} \end{pmatrix} \quad \begin{matrix} \text{level 1} \\ \\ \text{level 2} \end{matrix}$$

**Proposition 1.8** (retainment of decomposability/non-decomposability) *For any c-e structure $U$ there exists a two-level c-e structure $U' = \langle C_{U'}, E_{U'} \rangle$ such that $U$ is decomposable if and only if $U'$ is decomposable.* □

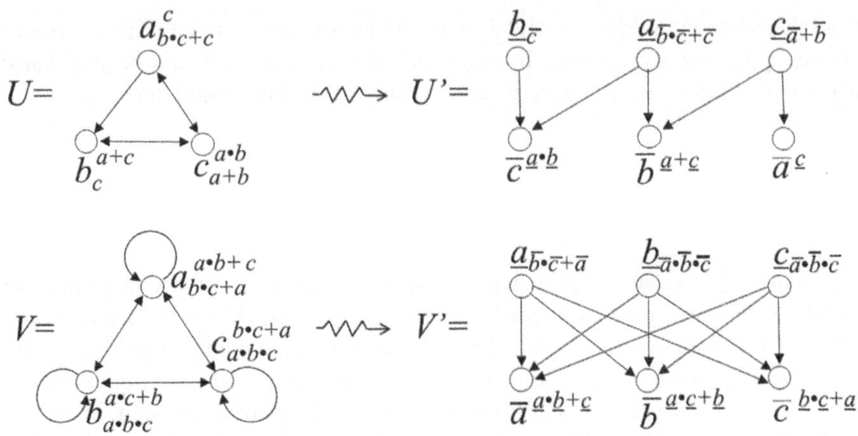


**Fig. 1.8** Transformation $U \rightsquigarrow U'$ where $U'$ is decomposable, thus also is $U$, and $V \rightsquigarrow V'$ where $V'$ is prime thus also is $V$

Transformation of c-e structures into two-level with the same feature pertaining to decomposition is exemplified in Fig. 1.8. The resulted two-level c-e structure is built-up of multiple replicas of nodes appearing in the original one. The replicas in the level 1 are underlined, while in level 2–with overbars.

Decomposability of $U'$ is obvious: its representation by arrow expression is:

$$(\underline{b} \to \overline{c}) \bullet ((\underline{a} \to \overline{c}) + (\underline{a} \to \overline{c}) \bullet (\underline{a} \to \overline{b})) + (\underline{c} \to \overline{b}) + (\underline{c} \to \overline{a})$$

Non-decomposability of $V'$ follows from the example next to Proposition 1.9: $V'$ becomes $U_3$ after appropriate change of names of nodes.

**Proposition 1.9** (infinite subset of prime c-e structures) *If the set* $\mathbb{X}$ *of nodes is infinite then there exist infinitely many prime c-e structures in the set* $\mathbf{CE}[\mathbb{X}]$. *Moreover, for each* $m \geq 6$ *there exists a prime two-level c-e structure* $U$ *with* $|car(U)| = m$. □

An example of infinite sequence of different non-decomposable c-e structures is $U_3, U_4, \ldots, U_n, \ldots$ where:

$$U_n = \begin{pmatrix} a^{\theta}_{\mathbf{1}+\mathbf{2}\bullet\mathbf{3}\bullet\ldots\bullet\mathbf{n}} & b^{\theta}_{\mathbf{1}\bullet\mathbf{2}\bullet\mathbf{3}\bullet\ldots\bullet\mathbf{n}} & c^{\theta}_{\mathbf{1}\bullet\mathbf{2}\bullet\mathbf{3}\bullet\ldots\bullet\mathbf{n}} & & \\ \mathbf{1}^{a\bullet b+c}_{\theta} & \mathbf{2}^{a\bullet c+b}_{\theta} & \mathbf{3}^{b\bullet c+a}_{\theta} & \mathbf{4}^{b\bullet c+a}_{\theta} & \ldots\ \mathbf{n}^{b\bullet c+a}_{\theta} \end{pmatrix} \begin{matrix} \text{level 1} \\ \text{level 2} \end{matrix}$$

for $n \geq 3$. $U_n$ is a two-level c-e structure: the level 1 consists of nodes $a, b, c$, the level 2 – of nodes $\mathbf{1}, \mathbf{2}, \mathbf{3}, \ldots, \mathbf{n}$ (boldface numerals are names of nodes). Let $U_n = \langle C, E \rangle$. Then $C(a) = C(b) = C(c) = E(\mathbf{j}) = \theta$ for $\mathbf{j} = \mathbf{1}, \mathbf{2}, \mathbf{3}, \ldots, \mathbf{n}$ and $C(\mathbf{j}) = b \bullet c + a$ for $\mathbf{j} = \mathbf{3}, \mathbf{4}, \ldots, \mathbf{n}$. One may demonstrate that $U_n$ is prime for all $n \geq 3$.

Obviously, any two-level c-e structure with a single node at one level is decomposable into single arrows. Indeed, such c-e structure with a single node e.g. at level 1 is of the form $U = \{a_{K(x1,x2,\ldots,xn)}, x1^a, x2^a, \ldots, xn^a\}$, where $K(x1, x2, \ldots, xn)$ is a polynomial of arguments $x1, x2, \ldots, xn$. It may be represented as $\overline{U} = \overline{K}(a \to x1,\ a \to x2, \ldots,\ a \to xn)$ where $\overline{K}$ is $K$ with each $x1, x2, \ldots, xn$ replaced with $a \to x1,\ a \to x2, \ldots,\ a \to xn$ respectively, and $+$, $\bullet$ meant as operations on c-e structures. But a somewhat unexpected property of c-e structures is stated in the next Proposition.

**Proposition 1.10** (small c-e structures are decomposable) *Any two-level c-e structure with* 3, 4 *or* 5 *nodes is decomposable into single arrows. The smallest prime is of 6 nodes.* □

Note that the decomposability/non-decomposability is a structural property, similarly as the structural deadlock, that is when $\boldsymbol{FC}[U] = \varnothing$. The decomposable c-e structure $\{a^{\theta}_{x+y}, b^{\theta}_{x\bullet y}, x^{a\bullet b}_{\theta}, y^{a\bullet b}_{\theta}\}$ and all prime c-e structures $U_n$ for $n \geq 3$ in above example are structural deadlocks. Another example is the decomposable c-e structure $\{a^{\theta}_{b\bullet c}, b^{a}_{c}, c^{a\bullet b}_{\theta}\}$ which is a structural deadlock. So, structural deadlocks may be either prime or decomposable. The converse question: is every prime c-e structure a structural deadlock, has negative answer. The counter-example is:

$$U = \begin{pmatrix} a^{\theta}_{x\bullet y+x\bullet z+u} & b^{\theta}_{x\bullet y+z\bullet u} & c^{\theta}_{x\bullet u+y\bullet z} & \\ x^{a\bullet b+b\bullet c}_{\theta} & y^{a\bullet b+c}_{\theta} & z^{a\bullet b\bullet c}_{\theta} & u^{a\bullet b\bullet c}_{\theta} \end{pmatrix}$$

Verification of its non-decomposability may be done by getting to contradiction: assuming $U = V \bullet W$, or $U = V + W$, one comes to conclusion that existence of $V$ and $W$ such that $\theta \neq V \neq U \neq W \neq \theta$ is impossible. Thus $U$ is prime. But $U$ is not a structural deadlock. Indeed, $U = \{a^{\theta}_{x\bullet y}, b^{\theta}_{x\bullet y}, x^{a\bullet b}_{\theta}, y^{a\bullet b}_{\theta}\} + U$, thus $\boldsymbol{FC}[U] = \{\{a^{\theta}_{x\bullet y}, b^{\theta}_{x\bullet y}, x^{a\bullet b}_{\theta}, y^{a\bullet b}_{\theta}\}\} \neq \varnothing$. Hence the Proposition 1.11.

**Proposition 1.11** (independent properties) *The notions of structural deadlock and decomposability of c-e structures are independent.* □

### *1.1.3 C*

The elementary c-e structures presented so far correspond, as shown later, to elementary nets [12], the primal version of Petri nets, expanded afterwards with external concepts like multi-valued places (with capacities) and arrow weights, what yielded the place/transition nets. For cellular version of c-e structures extension by the multi-valued places-cells only is adopted, the feature indispensable in cellular automata.[10]

It should be emphasized that in the abstract terms, the general c-e structures, as well as their cellular version and cellular automata are structurally the same. All are ordinary graphs, not bipartite like Petri nets, which structurally differ from the latter. The distinction lies in semantics, that is in the next-state rules. Essential feature of cellular c-e structures, imitating cellular automata, is a selection, i.e. adaptation of next-state rules for different modelled tasks, not a common for every task, as in case of general c-e structures and Petri nets. Less conceptually important but more convenient and general in some formulations, is representing the next-state rule of a c-e structure $U$ (and its cellular variant), as a relation $[\![U]\!]$ between pair of states. Not as a function, as in cellular automata and Petri nets. The extensions of elementary c-e structures,[11] apart from multivalence of nodes, involve arrows weighted with non-negative integers supplemented by $\omega$ which represents infinity, and where weight 0 does not affect value of nodes coincidental with the arrow ("read-only"),[12] whereas weight $\omega$ makes it inhibiting arrow.[13] So, weights 0 which make initial nodes of such arrows "read only", with multivalence of nodes will be concepts fundamental for cellular c-e structures. Let us quote from [5] these extensions, where multisets rather than sets like so far, are used in formal presentation of main concepts. The following notation for multisets will be convenient. Let $\mathbb{N}$ be the set of all natural numbers including 0 and let $\mathbb{N}_\omega = \mathbb{N} \cup \{\omega\}$, where the value $\omega$ symbolises infinity, that is $\omega > n$ for each $n \in \mathbb{N}$. A *multiset* over a base set $X$ is a (total) function $f : X \to \mathbb{N}_\omega$. If the set $\{x : f(x) \neq 0\}$ is finite then the linear-form notation is adopted for multisets, e.g.

| $X$ | $a$ | $b$ | $c$ | $d$ | $e$ |
|---|---|---|---|---|---|
| $f(X)$ | 2 | 0 | 3 | 1 | $\omega$ |

is denoted as $2 \otimes a + 3 \otimes c + d + \omega \otimes e$. A multiset is *zero* $\mathbb{O}$, when $\mathbb{O}(x) = 0$ for all $x$. Addition, subtraction and multiplication on multisets is defined as: $(f + g)(x) = f(x) + g(x)$, $(f - g)(x) = f(x) - g(x)$ for $f(x) \geq g(x)$, $(f \cdot g)(x) = f(x) \cdot g(x)$, comparison of multisets: $f \leq g$ iff $f(x) \leq g(x)$ for all $x$. Assume the

---

[10] In the case of place/transition Petri nets the arrows' weights must accompany multi-valued places, since these two concepts but jointly, constitute semantics contrived for an intended class of tasks (weights 1 are usually skipped in graphical presentation). This is not so with cellular c-e structures where the semantics uses the multivalence of cells only.

[11] Extension of c-e structures by time is not presented, as time cellular c-e structures are not introduced here.

[12] "Read only" arrows have been introduced by Winkowski [18], where they are named "read arcs", as a special kind of arrows, thus not by the extension of possible weights with integer number 0.

[13] The reason for introducing weight 0 and infinite weight $\omega$ in extended c-e structures is the following. Although a correct elementary c-e structure may contain a tight-cycle, like $\{x^x_{x \bullet y}, y^x_\theta\}$, it cannot pass a token from $x$ to $y$, and in the same action retaining it at $x$ (a structural deadlock – no firing components). But this is desirable and used in many examples as for instance in the c-e structure $\{x^x_{0 \otimes (x \bullet y)}, y^x_\theta\}$. Also, c-e structures lacking weights 0, cannot get rid of tokens unless supplemented by extra nodes gathering and removing tokens. Note that all this is possible in place/transition Petri nets (without any extras), where tight-cycles and "dangling" transitions show visible effects, that is a change of state.

customary arithmetic of $\omega$: $\omega + n = \omega$, $\omega - n = \omega$, $\omega + \omega = \omega$ and additionally $0 - \omega = 0$.

**Definition 1.10** (*state of c-e structures*) A state of extended c-e structure $U$ is a total function $s : car(U) \rightarrow \mathbb{N}$, thus a multiset over $car(U)$. The set of all states of $U$ is denoted by $\mathbb{S}$. □

**Definition 1.11** (*weights of monomials and capacity of nodes*) For a c-e structure $U = \langle C, E \rangle$ and its firing component $Q \in \boldsymbol{FC}[U]$, let with the pre-set ${}^{\bullet}Q$ and post-set $Q^{\bullet}$ of $Q$, some multisets $\overline{{}^{\bullet}Q}$: ${}^{\bullet}Q \rightarrow \mathbb{N}_{\omega}$ and $\overline{Q^{\bullet}}$: $Q^{\bullet} \rightarrow \mathbb{N}_{\omega}$ be given as additional information. The value $\overline{{}^{\bullet}Q}(x)$ is a *weight* (or *multiplicity*) of monomial $E_Q(x)$ and the value $\overline{Q^{\bullet}}(x)$ – a *weight* (or *multiplicity*) of monomial $C_Q(x)$. For $E_Q(x) = \theta$ and $C_Q(x) = \theta$ assume respectively $\overline{{}^{\bullet}Q}(x) = 0$ and $\overline{Q^{\bullet}}(x) = 0$. Additionally let $\overline{{}^{\bullet}Q}(x) = 0$ for $x \notin {}^{\bullet}Q$ and $\overline{Q^{\bullet}}(x) = 0$ for $x \notin Q^{\bullet}$. Let $cap(U)$ be a total function $cap(U) : car(U) \rightarrow \mathbb{N}_{\omega} \backslash \{0\}$, assigning a *capacity* to nodes in the set $car(U)$. A c-e structure endowed with such information is a *c-e structure-with-weighted monomials and capacity of nodes*. Note that this definition extends directly from the firing components onto their volleys. Indeed, for any non-empty volley $G \in \boldsymbol{FCV}[U]$ and any $x \in {}^{\bullet}G^{\bullet}$ there exists exactly one firing component $Q \in G$, with $x \in {}^{\bullet}Q^{\bullet}$ (because firing components in $G$ are pairwise detached). Thus one may define $\overline{{}^{\bullet}G} = \overline{{}^{\bullet}Q}$ and $\overline{G^{\bullet}} = \overline{Q^{\bullet}}$. It should also be noticed that weights of cause or effect monomials in identical firing components appearing in different c-e structures, may be different. □

An effect monomial $E_Q(a)$ of a node $a \in {}^{\bullet}Q$, with weight $\overline{{}^{\bullet}Q}(a)$, is denoted by $\overline{{}^{\bullet}Q}(a) \otimes E_Q(a)$. Similarly, a cause monomial $C_Q(x)$ of a node $x \in Q^{\bullet}$ with weight $\overline{Q^{\bullet}}(x)$ is denoted by $\overline{Q^{\bullet}}(x) \otimes C_Q(x)$. The coefficient representing weights will be abandoned if they are 1.

**Definition 1.12** (*inhibitors*) For a firing component $Q \in \boldsymbol{FC}[U]$, let $inh[Q] = \{x \in {}^{\bullet}Q : \overline{{}^{\bullet}Q}(x) = \omega\}$, thus a set of nodes in the pre-set of $Q$, whose effect monomials $E_Q(x)$ are of weight $\omega$. The nodes in $inh[Q]$ will play role of *inhibiting!node* of firing component $Q$. In accordance with Definition 1.5 (pairwise detached firing components in volleys), the concept of inhibiting nodes extends directly onto the volleys: $inh[G] = \{x \in {}^{\bullet}G : \overline{{}^{\bullet}G}(x) = \omega\}$ where $G \in \boldsymbol{FCV}[U]$. The inhibiting nodes will be called *inhibitors*. □

**Definition 1.13** (*enabled firing components and enabled volleys*) For a firing component $Q \in \boldsymbol{FC}[U]$ and state $s$ let us define the formula: *enabled*$[Q](s)$ if and only if:

$$\forall x \in inh[Q] : s(x) = 0 \ \wedge$$
$$\forall x \in {}^{\bullet}Q \setminus inh[Q] : s(x) > 0 \wedge \overline{{}^{\bullet}Q}(x) \leq s(x) \leq cap(U)(x) \ \wedge$$
$$\forall x \in Q^{\bullet} : \overline{Q^{\bullet}}(x) + s(x) \leq cap(U)(x)$$ □

So, $Q$ is enabled in the state $s$ iff none of inhibiting nodes $x \in {}^\bullet Q$ contains a token and each remaining node in ${}^\bullet Q$ contains, with no fewer tokens than is the weight of its effect monomial $E_Q(x)$ and no more than capacity of each $x \in {}^\bullet Q$. Moreover, none of $x \in Q^\bullet$ holds more tokens than their number, when increased by the weight of the cause monomial $C_Q(x)$, exceeds capacity of $x$. Evidently, for a volley $G \in \boldsymbol{FCV}[U]$, the formula $enabled[G](s)$ is defined as above by replacing $Q$ with $G$ and $\boldsymbol{FC}[U]$ with $\boldsymbol{FCV}[U]$.

**Definition 1.14** (*sequential and parallel semantics of extended c-e structures*)

**Sequential** For $Q \in \boldsymbol{FC}[U]$, let $[\![Q]\!] \subseteq \mathbb{S} \times \mathbb{S}$ be a binary relation defined as:

$$(s, t) \in [\![Q]\!] \text{ iff } enabled[Q](s) \wedge t = (s - \overline{{}^\bullet Q}) + \overline{Q^\bullet} \leq cap(U)$$

($Q$ transforms state $s$ into $t$). *Semantics* $[\![U]\!]$ of $U \in \boldsymbol{CE}$ is $[\![U]\!] = \bigcup_{Q \in \boldsymbol{FC}[U]} [\![Q]\!]$.

Closure $[\![U]\!]^*$ and reachability and computation are defined as in Definition 1.7.

**Parallel** For a volley $G \in \boldsymbol{FCV}[U]$, $G \neq \varnothing$, the relations $[\![G]\!]$ and $[\![U]\!]_{par}$ are defined in the same way as $[\![Q]\!]$ and $[\![U]\!]$ in the sequential case, but with $Q$ replaced with $G$ and $\boldsymbol{FC}[U]$ replaced with $\boldsymbol{FCV}[U]$. Closure $[\![U]\!]^*_{par}$, reachability and computation are defined as in the sequential case. □

An example of using inhibitor, i.e. node from which an arrow weighted by $\omega$ emanates, as well as an arrow weighted by 0 ("read only") is in Fig. 1.9. The weight of one effect (i.e. subscript) monomial $\omega \otimes \mathbf{z}$ of the node $\mathbf{T}$ is $\omega$, thus traversing arrow from $\mathbf{T}$ to $\mathbf{z}$ would require infinite number of tokens at $\mathbf{T}$, which is impossible. Thus, $\mathbf{T}$ is the inhibiting node in firing component $Q_0$ and ordinary node – in firing component $Q_1$. In accordance with notation of multisets, the weighted monomials are denoted by $\mathrm{w} \otimes M$, where w is the weight of $M$ – an unweighted monomial. The weight is skipped if $\mathrm{w} = 1$. The corresponding arrows are red and dashed in the figure if $\mathrm{w} = \omega$.

In example shown in Fig. 1.9, the operation on c-e structures with capacity of places and weights, also 0 ("read only") and $\omega$ (inhibiting) has been applied. In general, operations (+), (•) on extended c-e structures can be applied under some

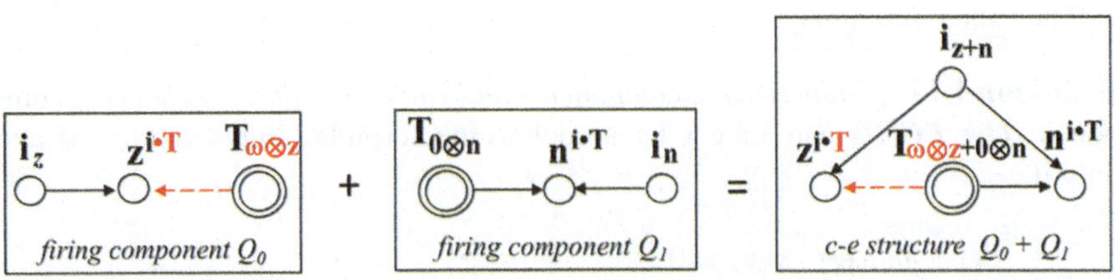


**Fig. 1.9** C-e structure $Q_0 + Q_1$ implementing "test zero". A token at the node $\mathbf{i}$ starts testing contents of $\mathbf{T}$; on termination, the token appears at $\mathbf{z}$ (zero) if $\mathbf{T}$ is empty and at $\mathbf{n}$ (not zero) otherwise. Node $\mathbf{T}$ plays two roles: inhibiting in $Q_0$ and "read only" in $Q_1$. Capacity of $\mathbf{T}$ is infinite, its content remains unchanged on termination due to the monomial $0 \otimes \mathbf{n}$

conditions only. Such conditions should secure avoidance of ambiguities in resulted c-e structures, caused by different weights of identical arrows in particular constituents of an operation, or different capacities of identical nodes. Such conditions are fulfilled in case of example in Fig. 1.9. Not entering this issue in detail, one can notice that combining extended c-e structures would require to redefine the operations $(+)$, $(\bullet)$ to be partial, not total as in Definition 1.3. Thus, the calculus of c-e structures would become a partial algebra (cf [6]). Instead, it is assumed that along with a given elementary c-e structure $U \in \boldsymbol{CE}[\mathbb{X}]$ (before then constructed by means of operations introduced in Definition 1.3) and the set of firing components $\boldsymbol{FC}[U] = \boldsymbol{SUB}[U] \cap \boldsymbol{FC}$, the additional features provided in Definitions 1.10, 1.11 are associated. Two of these features will be indispensable for cellular version of c-e structures: infinite capacity of nodes-cells and weight 0 of arrows. Note that endowments of c-e structures with read only and inhibiting capabilities has been done by taking the set of all natural numbers with 0 and $\omega$ as weights, not mere positive integers, like e.g. in place/transition Petri nets. So, not by introducing new structural entities i.e. a special kind of arrows.[14]

Concluding this outline of c-e structures as an introductory and the wherewithal for the main topic of the book, let us take a look at some equivalences between them and Petri nets. It may seem that reciprocal conversion between c-e structures and Petri nets, preserving behaviour, is straightforward: the nodes play part of net places whereas firing components–transitions. Such simple one to one, i.e. bijective correspondence may be established from c-e structures to nets but not always from nets to c-e structures, if "one to one" concerns relationship between nodes of c-e structure's carrier and places of Petri net as well as between firing components and transitions. Nonetheless behavioural equivalence between c-e structures and Petri nets is possible, if a slight modification of the c-e structures is made–when needed.[15] The reciprocal (two-way) conversion: c-e structures $\leftrightsquigarrow$ Petri nets, in their elementary versions, employs the addition $(+)$ and multiplication $(\bullet)$, as will be shown in Definition 1.16. These operations are total, i.e. do not require any restrictions on arguments, so they operate with means the algebra of c-e structures offers. This is not so with extended c-e structures, where combining c-e structures requires that the same arrows and nodes appearing in the c-e structures being combined, have the same weights and capacities. This needs partial operations, i.e. partial algebra (cf [6]). Here, let us limit this problem to the elementary c-e structures and Petri nets, let us recall the definition of the latter, and define two kinds of equivalences between them and c-e structures.

A net is a triple $\mathcal{N} = \langle S, T, F \rangle$ where $S$ is a set of places, $T$ – a set of transitions with $S \cap T = \varnothing$, and $F \subseteq S \times T \cup T \times S$ set of directed edges, called a flow relation. A *pre-set* and *post-set* of a transition $t \in T$ is ${}^{\bullet}t = \{x \in S : (x, t) \in F\}$ and

[14] Remember that in c-e structures there is no separate formal notion termed "arrow". The phrase "weight of arrow" means the coefficient $k$ in monomials annotating nodes, like $k \otimes y$ in "arrow expression" $\{x^{\theta}_{k\otimes y},\ y^{x}_{\theta}\}$ informally denoted and depicted as $x \xrightarrow{k} y$ where $k \in \mathbb{N}_{\omega}$.

[15] The first publication on non convertibility of each elementary Petri net into elementary c-e structures appeared in Raczunas [14].

$t^\bullet = \{x \in S : (t, x) \in F\}$ respectively and dually for places $x \in S$: $t \in {}^\bullet x \Leftrightarrow x \in t^\bullet$ and $t \in x^\bullet \Leftrightarrow x \in {}^\bullet t$. To conform to the construction of c-e structures, let us admit (not essential) restrictions: ${}^\bullet t \neq \varnothing \neq t^\bullet$ and $t_1 \neq t_2 \Rightarrow ({}^\bullet t_1 \neq {}^\bullet t_2 \vee t_1^\bullet \neq t_2^\bullet)$ for $t, t_1, t_2 \in T$. The latter means that we consider the so-called simple nets only. A marking $M$ of a net is a counterpart of the c-e structures' state and firing rule – a counterpart of semantics: $M \subseteq S$ and $(M, M') \in [\![t]\!]$ iff ${}^\bullet t \subseteq M \wedge t^\bullet \cap M = \varnothing \wedge M' = (M \backslash {}^\bullet t) \cup t^\bullet$ for $t \in T$ and $[\![\mathcal{N}]\!] = \bigcup_{t \in T} [\![t]\!]$.

**Definition 1.15** (*strong equivalence*) A net $\mathcal{N} = \langle S, T, F\rangle$ and a c-e structure $U = \langle C, E\rangle$ are *strongly equivalent* if and only if there exist two bijections $g : S \to car(U)$ and $f : T \to \boldsymbol{FC}[U]$ such that $g({}^\bullet t) = {}^\bullet f(t)$ and $g(t^\bullet) = f(t)^\bullet$ for each $t \in T$, where $g({}^\bullet t) = \{g(x) : x \in {}^\bullet t\}$, $g(t^\bullet) = \{g(x) : x \in t^\bullet\}$. □

Obviously, ${}^\bullet f(t)$ and $f(t)^\bullet$ are pre and post sets of the firing component $f(t)$–see Definition 1.4, while ${}^\bullet t$ and $t^\bullet$ are pre and post sets of the transition $t$. This structural equivalence yields the identical behaviour: if a net $\mathcal{N}$ and c-e structure $U$ are strongly equivalent then $[\![\mathcal{N}]\!]$ and $[\![U]\!]$ are identical up to naming of places in $\mathcal{N}$ and nodes in $U$. But as will be seen, the requirement that $g$ is bijective turns out too strong. For behavioural equivalence of $\mathcal{N}$ and $U$, bijection $f$ only is necessary.

**Definition 1.16** (*direct conversion*) Suppose that names (labels) of net places and c-e structure nodes are taken from the same universe $\mathbb{X}$. For a given c-e structure $U = \langle C, E\rangle$ define a net $\mathcal{N} = \langle S, T, F\rangle$ as follows: $S = car(U)$, $T = \boldsymbol{FC}[U]$, $F = \{(x, Q) : x \in {}^\bullet Q \wedge Q \in T\} \cup \{(Q, x) : x \in Q^\bullet \wedge Q \in T\}$. For a given net $\mathcal{N} = \langle S, T, F\rangle$ define a c-e structure $U = \langle C, E\rangle$ as follows:

$$C(x) = \sum_{t \in T_x} \bigwedge_{y \in {}^\bullet t} y \qquad \text{where } T_x = \{t \in T : x \in t^\bullet\}$$

$$E(x) = \sum_{t \in T^x} \bigwedge_{y \in t^\bullet} y \qquad \text{where } T^x = \{t \in T : x \in {}^\bullet t\}$$

"$\bigwedge$" stands for the operator of repetitive multiplication "$\bullet$" of c-e structures. These conversions are called *direct*. □

Figure 1.10 depicts an example of direct reciprocal conversion $\leftrightsquigarrow$ between c-e structure $U$ and Petri net $\mathcal{N}$. This conversion yields the strong equivalence of $U$ and $\mathcal{N}$: the number of nodes in $U$ and places in $\mathcal{N}$ is the same, as well as the number of firing components in $U$ and transitions in $\mathcal{N}$. The shaded boxes in $\mathcal{N}$, that represent transitions, are the firing components of $U$.

As said above, direct conversion of c-e structures into Petri nets yields their strong equivalence thus preserves behaviour. This always is possible: the mappings $g(x) = x$ and $f(Q) = Q$, for $x \in car(U)$, $Q \in \boldsymbol{FC}[U]$ satisfy requirements of Definition 1.15 when the net $\mathcal{N}$ is constructed from $U$ by the direct conversion. The direct conversion of a Petri net into a c-e structure can sometimes yield strongly equivalent c-e structure, but sometimes this is not possible, as shows Fig. 1.11. The

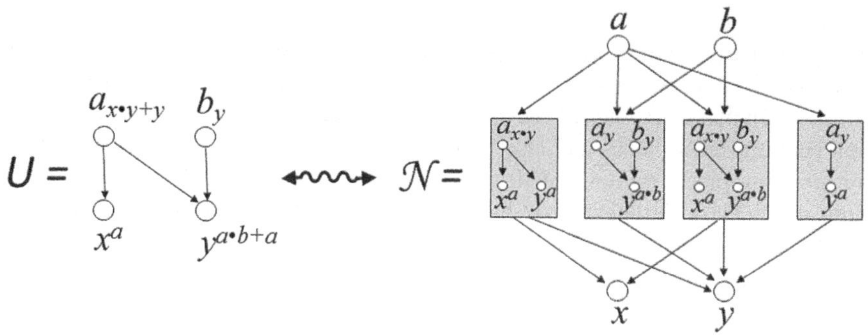


**Fig. 1.10** Strongly equivalent c-e structure $U = \{a_{x\bullet y+y}, b_y, x^a, y^{a\bullet b+a}\}$ with Petri net $\mathcal{N}$. Both behave identically

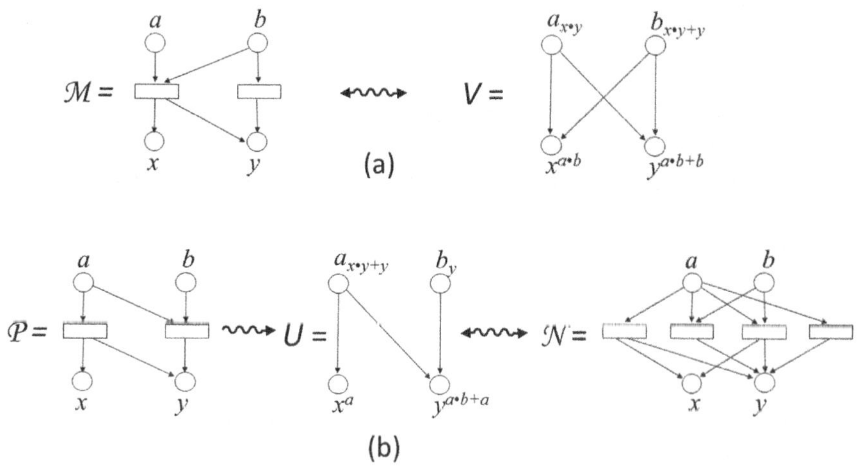


**Fig. 1.11** **a** Direct reciprocal conversion $\mathcal{M} \leftrightsquigarrow V$ yields strong equivalence, thus $[\![\mathcal{M}]\!] = [\![V]\!]$; **b** direct one-way conversion $\mathcal{P} \rightsquigarrow U$ does not yield strong equivalence, but direct two ways $U \leftrightsquigarrow \mathcal{N}$– does. Therefore $[\![\mathcal{P}]\!] \neq [\![U]\!] = [\![\mathcal{N}]\!]$

case (a) shows direct conversions of Petri net $\mathcal{M}$ into c-e structure $V$ and $V$ into $\mathcal{M}$. Both conversions yield strong equivalence of $\mathcal{M}$ and $V$, thus ensure their identical behaviour. The case (b) shows direct conversion of Petri net $\mathcal{P}$ into c-e structure $U$, not being strongly equivalent to $\mathcal{P}$, since this conversion creates more firing components than $\mathcal{P}$ has transitions. Thus $\mathcal{P}$ and $V$ would behave differently. Direct conversion of $U$ yields strongly equivalent Petri net $\mathcal{N}$ but not $\mathcal{P}$, what has been shown in Fig. 1.10.

The reason for non-existence of a strongly equivalent c-e structure for each Petri net is the creation of superfluous (redundant) firing components in summation–recall point 10 in Proposition 1.2 and point 7 in Proposition 1.3. Although not always it is possible to construct a strongly equivalent c-e structure for a given Petri net, the strong equivalence, as too restrictive, can be relaxed, nonetheless preserving their behavioural equivalence. Such weakened equivalence ensures bijective correspondence between transitions in Petri nets and firing components in c-e structures but not between places and nodes. To this end, we define an auxiliary equivalence relation between net's places and also between c-e structure's nodes. This will allow for grouping places and also nodes into equivalence classes. All members of any equivalence class simultaneously hold or do not hold tokens during activity of the c-e structure constructed from a given net. The weak equivalence will be defined by means of these equivalence classes.

**Definition 1.17** (*equivalence "$\equiv$" amid Petri net places; also amid c-e structure nodes*) Let a net $\mathcal{N} = \langle S, T, F\rangle$ and a c-e structure $U$ be given. For $x, y \in S$ or $x, y \in car(U)$ define $x \equiv y$ if and only if ${}^{\bullet}x = {}^{\bullet}y$ and $x^{\bullet} = y^{\bullet}$. Obviously $\equiv$ is an equivalence relation. Let us shorten the denotation of $\equiv$-equivalence classes $[x]_{\equiv}$ to $[x]$, and for a set $A \subseteq S$ or $A \subseteq car(U)$ let us denote the set $\{[x] : x \in A\}$ by $[A]$. □

At each marking (state) reachable from an initial one, all the places (nodes) in the same equivalence class, either simultaneously hold tokens or do not hold tokens – provided that this is so at the initial marking (state). $[A]$ is the quotient set (with respect to $\equiv$) of $A$.

**Definition 1.18** (*weak equivalence*) A net $\mathcal{N} = \langle S, T, F\rangle$ and a c-e structure $U = \langle C, E\rangle$ are *weakly equivalent* if and only if there are two bijections $g : [S] \to [car(U)]$ and $f : T \to \boldsymbol{FC}[U]$ such that $g([{}^{\bullet}t]) = [{}^{\bullet}f(t)]$ and $g([t^{\bullet}]) = [f(t)^{\bullet}]$ for each $t \in T$. □

Thus, the difference between strong and weak equivalence concerns the bijection $g$ only: in the strong equivalence its domain and codomain are sets $S$ and $car(U)$, whereas in the weak equivalence – quotients $[S]_{\equiv}$ and $[car(U)]_{\equiv}$. It is seen that this definition ensures the property expressed above. That is, a transition $t \in T$ and firing component $f(t)$ is enabled iff all places $x \in \bigcup[{}^{\bullet}t]$ and nodes $g(x)$ hold tokens and all places $y \in \bigcup[t^{\bullet}]$ and nodes $g(y)$ – do not.

Figure 1.12 depicts an example of different than direct conversion of the Petri net into a weakly equivalent c-e structure.

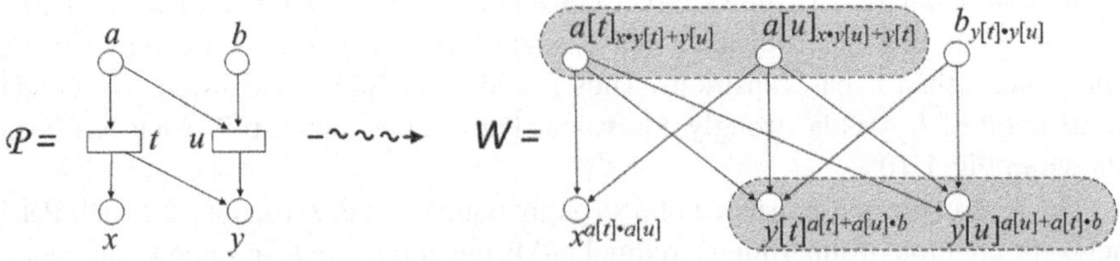


**Fig. 1.12** Petri net $\mathcal{P}$ converted into weakly equivalent c-e structure $W$. Groups of the replicas of $a$ and $y$ are encircled and shaded. Here $[\![\mathcal{P}]\!] = [\![W]\!]$

Summarizing these notes on relationship between c-e structures and Petri nets, notice that direct conversion between them, was made by two simple procedures expressed in Definition 1.16. Applied to c-e structures, they always lead to strongly equivalent Petri nets, so behaving identically, though not always the other way round. But identical behaviour is obtained only when bijective correspondence between transitions and firing components is secured and when similar conditions make them enabled. In order to get always identical behaviour of c-e structures and Petri nets, another than the direct procedure of Petri nets conversion into c-e structures has to be applied. Such conversion procedure, which leads to weakly equivalent c-e structures is much more complicated than this in Definition 1.16 and is described in detail in [5, 6]. Its essence consists in recognizing and eliminating extraneous (not corresponding to any Petri net transition) firing components appearing in summation applied in Definition 1.16. Finally, notice that the cellular c-e structures, the main subject of this book, structurally a subclass of general c-e structures, resemble the so-called "state machines", a sub-class of Petri nets, because of non-existence of multiplication.

## 1.2 Fix-Point Framework

The origins of the elementary cause-effect structures date back to the mid 1980, when several texts on this subject appeared as internal reports of some universities or proceedings of conferences, then, like [2 4, 8, 14, 17] in information science periodicals. All of them presented elementary c-e structures as directed graphs in which predecessors and successors of each vertex are grouped into its cause and effect regions, like those encircled in Fig. 1.3a. In the fix-point approach the vertices, called network nodes, assume status "active" or "passive", formally treated as logical values ***true*** or ***false***, denoted **1**, **0**, while visually perceived as a presence or absence of a token representing a single signal. The activity of such structure consists in dispatch a signal to a node, simultaneously from all nodes forming one of its cause region, and symmetrically, in reception of a signal from a node, simultaneously by all nodes forming one of its effect region. The choice of a region in such action is indeterministic. The active node is <u>ready</u> to send signal to one of its effect regions, whereas the passive is <u>ready</u> to receive signals from one of its cause regions. But the readiness does not mean the node's capability to perform such action. This depends on analogous capability of nodes forming its cause or effect regions. Such interrelated conditions may be expressed as follows. A node <u>can</u> become active if it is passive and each of its predecessors from at least one cause region <u>can</u> become passive. And symmetrically: a node <u>can</u> become passive if it is active and each of its successors from at least one effect region <u>can</u> become active. This recursive phrase, expressing a symmetry between dispatch and reception, leads to equations whose solution defines the semantics of the cause-effect structure. Their solution determines when a node $x$ can change its status, active to passive or conversely, and yields a family of the so-called *impact regions* of $x$, i.e. sets of nodes whose status change should occur along with status change of $x$. Only one such impact region (a firing component introduced in Definition 1.4), is chosen "to fire". The equations for a c-e structure $U = \langle C, E\rangle$ are functional in that each node $x \in car(U)$ is associated with two Boolean functions of state,[16] one for reception: $x^{C(x)} : \mathbb{S} \to \{\mathbf{1}, \mathbf{0}\}$ and one for dispatch: $x_{E(x)} : \mathbb{S} \to \{\mathbf{1}, \mathbf{0}\}$, returning $\mathbf{1}$, if $x$ can change its status in a given state and $\mathbf{0}$ otherwise,[17] and called a *cause trigger* and *effect trigger* of the node $x$. Assuming that polynomials $C(x)$, $E(x)$ are sums of monomials (Definition 1.3): $C(x) = P + \ldots + Q$ and $E(x) = R + \ldots + S$, the triggers are defined as disjunctions of functions: $x^{C(x)} = \overline{x} \wedge (x^P \vee \ldots \vee x^Q)$ and $x_{E(x)} = x \wedge (x_R \vee \ldots \vee x_S)$, where Boolean functions $x^P, \ldots, x^Q$, and $x_R, \ldots, x_S$ are the least[18] solutions to functional equations straightforwardly read out from the c-e structure $U$. So, these functions are unknowns of the equations, i.e. sought-after, whereas any node $x$, regarded as a Boolean function of state defined by $x(s) = s(x)$,[19] is the known given parameter. The solutions to these equations are of the following general form, where $\overline{x} = \neg x$ and a letter with apostrophe denotes either this letter itself (i.e. a known function) or its negation, so the letter with the overbar:

$$x^P = \overline{x} \wedge [(a' \wedge \ldots \wedge b') \vee \ldots \vee (c' \wedge \ldots \wedge d')]$$

$$\cdots\cdots\cdots\cdots\cdots\cdots\cdots\cdots\cdots$$

$$x^Q = \overline{x} \wedge [(e' \wedge \ldots \wedge f') \vee \ldots \vee (g' \wedge \ldots \wedge h')]$$

$$x_R = x \wedge [(p' \wedge \ldots \wedge q') \vee \ldots \vee (r' \wedge \ldots \wedge t')]$$

$$\cdots\cdots\cdots\cdots\cdots\cdots\cdots\cdots\cdots$$

$$x_S = x \wedge [(u' \wedge \ldots \wedge v') \vee \ldots \vee (y' \wedge \ldots \wedge z')]$$

Instead of going into details of this (rather mechanical) procedure of creating and solving equation system

$$x^{C(x)} = \overline{x} \wedge (x^P \vee \ldots \vee x^Q) \quad x_{E(x)} = x \wedge (x_R \vee \ldots \vee x_S) \quad \text{for all } x \in car(U)$$

for a given c-e structure $U$ (cf [2, 3]), let us illustrate it by example of c-e structure shown in Fig. 1.13.

---

[16] The Boolean functions of state are composed as usually: $[f \vee g](s) = f(s) \vee g(s)$, $[f \wedge g](s) = f(s) \wedge g(s)$, $\overline{f}(s) = \neg f(s)$, but the partial order between them is defined inversely: $f \leq g$ iff $\forall s \in \mathbb{S} : g(s) = \mathbf{1} \Rightarrow f(s) = \mathbf{1}$. So, zero $\bot$ and unit $\top$ functions are such that $\bot(s) = \mathbf{1}$ and $\top(s) = \mathbf{0}$ for all $s \in \mathbb{S}$, and the least solution to functional equations is with respect to this order.

[17] In Sect. 1.1, the symbols "$x^{C(x)}$", "$x_{E(x)}$", have been used to point out how node $x$ is interconnected with nodes appearing in formal polynomials $C(x)$, $E(x)$, so they referred to structure only. Here, apart from that, they play also semantical role, ensued from interconnection of $x$ with other nodes.

[18] See footnote 16.

[19] Such notation is used here and there, e.g. in [13].

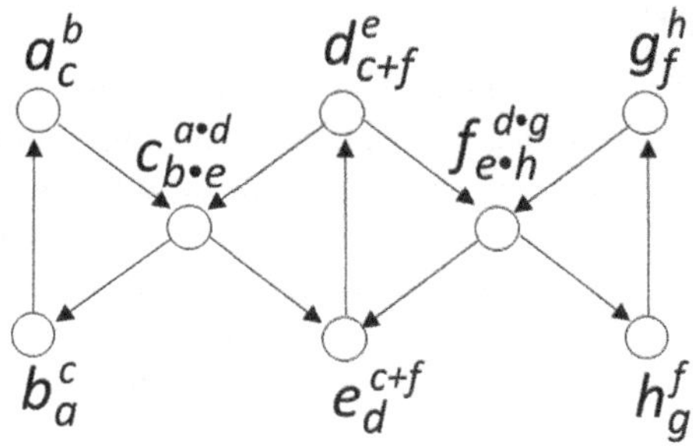


**Fig. 1.13** C-e structure whose activity determine equations written below

The Boolean functional equations for c-e structure in Fig. 1.13 are the following:
Reception of signal:

$$a^b = \overline{a} \wedge b_a \quad b^c = \overline{b} \wedge c_{b\bullet e} \qquad c^{a\bullet d} = \overline{c} \wedge a_c \wedge d_c$$

$$d^e = \overline{d} \wedge e_d \quad e^{c+f} = \overline{e} \wedge (c_{b\bullet e} \vee f_{e\bullet h}) \quad f^{d\bullet g} = \overline{f} \wedge d_f \wedge g_f$$

$$g^h = \overline{g} \wedge h_g \quad h^f = \overline{h} \wedge f_{e\bullet h}$$

Dispatch of signal:

$$a_c = a \wedge c^{a\bullet d} \qquad b_a = b \wedge a^b \quad c_{b\bullet e} = c \wedge b^c \wedge e^c$$

$$d_{c+f} = d \wedge [c^{a\bullet d} \vee f^{d\bullet g}] \quad e_d = e \wedge d^e \quad f_{e\bullet h} = f \wedge e^f \wedge h^f$$

$$g_f = g \wedge f^{d\bullet g} \qquad h_g = h \wedge g^h$$

The least solutions to these equations are:
Triggers for reception of signal:

$$a^b = \overline{a} \wedge b \quad b^c = \overline{b} \wedge c \wedge \overline{e} \qquad c^{a\bullet d} = \overline{c} \wedge a \wedge d$$

$$d^e = \overline{d} \wedge e \quad e^{c+f} = \overline{e} \wedge (c \vee f) \quad f^{d\bullet g} = \overline{f} \wedge d \wedge g$$

$$g^h = \overline{g} \wedge h \quad h^f = \overline{h} \wedge f \wedge \overline{e}$$

Triggers for dispatch of signal:

$$a_c = a \wedge d \wedge \overline{c} \qquad b_a = b \wedge \overline{a} \quad c_{b\bullet e} = c \wedge \overline{b} \wedge \overline{e}$$

$$d_{c+f} = d \wedge [(a \wedge \overline{c}) \vee (g \wedge \overline{f})] \quad e_d = e \wedge \overline{d} \quad f_{e\bullet h} = f \wedge \overline{e} \wedge \overline{h}$$

$$g_f = g \wedge d \wedge \overline{f} \qquad h_g = h \wedge \overline{g}$$

**Remarks**

1. The equations directly obtained from a c-e structure are being solved by the schoollike elimination of unknowns, in case of finite number of equations. But there in no need to confine ourselves to the finite systems, since the triggers, i.e. Boolean functions $x^{C(x)}, x_{E(x)}$, exist in the complete partial order (cpo) $\leq$ between functions of the type $\mathbb{S} \to \{\mathbf{1}, \mathbf{0}\}$ (see footnote 16), also for infinite number of equations. This is so because the operators $\wedge$, $\vee$ are continuous and the overbar "‾" is applied only to functions represented by names of nodes $x, y, \ldots$, thus it is not an operator in this cpo, but indicates that $\overline{x}(s)$ assumes value inverse to $x(s)$.
2. The essentially different approach to semantics of c-e structures in their axiomatic and fix-point framework should be noticed. In the former case the semantics of $U$ is a binary relation $[\![U]\!]$ between states, while in the latter – ascertainment whether or not a node in a given state can change its status into inverse. This difference originates in a certain dissent between the lattice of c-e structures and the lattice of Boolean functions. The lattice $\langle \boldsymbol{CE}, +, \circ \rangle$ of c-e structures (see Propositions 1.5, 1.6, 1.7) enjoys only the conditional distributivity of "$\circ$" over "+", whereas the lattice of Boolean functions with "$\vee$" and "$\wedge$" (see footnote 16)–the full distributivity of "$\wedge$" over "$\vee$". The former is non-distributive and incomplete, in contrast to the latter. This causes, among other things, that the lattice of c-e structures $\langle \boldsymbol{CE}, +, \circ \rangle$ as well as the quasi-semiring $\langle \boldsymbol{CE}, +, \bullet \rangle$ do not enjoy the absorptive capability: $a \circ b + b \neq b$, and $a \bullet b + b \neq b$, whereas the lattice of Boolean functions enjoys it: $(a \wedge b) \vee b = b$. This difference between two paradigms of semantics is of no matter for the cellular version of c-e structures, where only operator "+" is utilized.

## References


1. Bruns G.: Distributed Systems Analysis with CCS. Prentice-Hall (1997)
2. Czaja, L.: Cause-effect structures. Inf. Process. Lett. **26**(1987/88) 313–318. (North Holland)
3. Czaja, L.: Finite processes in cause-effect structures and their composition. Inf. Process. Lett. 31, June 1989. (North Holland)
4. Czaja, L.: A calculus of nets. Cybern. Syst. Anal. **29**(2) (1993)
5. Czaja, L.: Cause-Effect Structures. An Algebra of Nets with Examples of Application. Lecture Notes in Networks and Systems, vol. 45. Springer (2019)
6. Czaja, L.: Computing in Cause-Effect, Structures. Lecture Notes in Networks and Systems, vol. 331. Springer (2022)
7. Hoare, C.A.R.: Communicating Sequential Processes. Prentice Hall (1985)
8. Holenderski, L., Szalas, A.: Propositional description of finite cause-effect structures. Inf. Proc. Lett. **27**, 111–117 (1988). (North Holland)
9. Kari, J.J.: Cellular automata. In: Handbook of Natural Computing, vol. 4. Springer (2012)
10. Milner, R.: Communication and Concurrency. Prentice Hall Inc. (1989)
11. Von Neumann, J.: Theory of Self-Reproducing Automata. Published by University of Illinois Press, Urbana, IL (1966). (Edited and completed by Arthur W. Burks)
12. Petri, C.A.: Communication with automata. Final report, vol. 1, Supplement 1, RADC TR-65-337-vol. 1-suppl. 1, Applied Research, Princeton, NJ. Contract AF 30(602)-3324 (January 1966)

13. Rasiowa, H., Sikorski, R.: The Mathematics of Metamathematics. PWN–Polish Scientific Publishers (1968)
14. Raczunas, M.: Remarks on the equivalence of c-e structures and Petri nets. Inf. Proc. Lett. **45**, 165–169 (1993)
15. Reisig, W.: Petri Nets. An Introduction, Number 4 in EATCS Monographs on Theoretical Computer Science. Springer, Berlin-Heidelberg-New York, Tokyo (1985)
16. Roscoe, A.W.: The Theory and Practice of Concurrency. Prentice-Hall (1998)
17. Ustimenko, A.P.: Algebra of two-level cause-effect structures. Inf. Process. Lett. **59**, 325–330 (1996)
18. Winkowski, J.: Reachability in Contextual Nets. Fundamenta Informaticae **51**(1). (2, June 2002)
19. Wolfram, S.: Universality and complexity in cellular automata. Phys. D **10**, 1–35 (1984)

# Chapter 2
# Cellular Version of Cause-Effect Structures

## 2.1 Outline with examples

The key idea to exploit some concepts from c-e structures in constructing cellular c-e structures is to restrict operations on nodes and on c-e structures to addition (+), thus to neglect multiplication (•), and to apply only parallel semantics, which may be arbitrary binary relation between states. Thus, a relation specific to a modelled task. A consequence is to reduce notion of firing component (Chap. 1) to two nodes: one cause and one effect–on structural side–and to provide a rule of the next-state, separately for different modelled tasks–on semantic side. Not a rule common to all tasks as in case of standard c-e structures. So, now addition will not mean indeterministic choice, but only a structural constructor. That is, a syntactic operator combining nodes or c-e structures, in accordance with the following axioms:

$$\left.\begin{array}{l} \theta + U = U + \theta = U \\ U + U = U \\ U + V = V + U \\ U + (V + W) = (U + V) + W \end{array}\right\} \qquad (2.1)$$

where each $U$, $V$, $W$ stands either for a set of nodes, or a c-e structure. Thus, this reduced algebra of c-e structures is a commutative and idempotent monoid, not a general quasi semiring. The constructor "+", for sets of nodes, becomes just the union of sets. Any node will be called a *cell*. It is connected by bidirectional arrow with some other cells – its *neighbours*. Thus, each neighbour of any cell is both its cause and effect. Each cell has the same number of neighbours, which make its *neighbourhood*. Since the characteristic behavioural feature of cellular c-e structures is their potential of simultaneous impact on each cell when passing from a state to next-state, the parallel semantics is adopted. This feature supercedes semantic role of multiplication (•) in the c-e structures. Summarizing this sketchy presentation: *a cellular c-e structure is a c-e structure with cause and effect polynomials of each node*

 
L. Czaja, *Cellular Cause-Effect Structures*, Studies in Systems, Decision and Control 526,
https://doi.org/10.1007/978-3-031-55590-9_2

*(cell) being identical monomials (i.e. without multiplication), with equinumerous neighbourhoods of each cell and the next-state relation adapted to a particular task.*[1] But a peculiarity of cellular automata is their location in some geometric spaces and notably shape of cells and their ordered neighbourhood, the features absent in general c-e structures. This aspect will find formal regard later in constructing their cellular version. In particular a fixed order and orientation of neighbours around cells when combining cellular c-e structures. Also before sections where such feature is formally required, mostly the same order and orientation of cells in every neighbourhood is adopted in examples – for better readability. In such terms, all "atomic" structures – cells along with their neighbours, are structurally replicas (copies) of one another. All are geometrically isomorphic with respect to a shift of their location in a space, but not transformations like rotation or mirror image. That is, differing only by location on a base being a Cartesian space (of arbitrary dimension – in further chapters). In other words, the placement of cells relative to one another onto the base, will be identical in each atomic structure of a given cellular c-e structure. This is the only supplementary feature imposed on the c-e structures (devoid of multiplication) to obtain their cellular version, needed when cellular c-e structures are specified as sums of their atoms and when cellular c-e structures are being combined – the issue considered later.

Before getting into general cellular c-e structures, their planar (in geometric presentation) case with two possible contents ("status") of any cell–***true*** or ***false***–will be considered. Let us illustrate them on a typical example, often appearing in writings on cellular automata, cf e.g. [2, 4, 5].

## 2.1.1 Example (Game of Life)

### 2.1.1.1 The Game of von Neumann Type

The game takes place on the infinite 2-dimensional board being a grid of rectangular cells ("tiles"). Each cell, i.e. a node of a cellular c-e structure, is annotated by its neighbourhood, as shows Fig. 2.1. The nodes are named $a_{xy}$ $(-\infty < x, y < +\infty)$ where indices $x$, $y$ denote values of Cartesian coordinates identifying the cells on the grid.[2] The causes and effects of each node coincide:

---

[1] The cellular c-e structures work like special parallel deterministic automata without external inputs. Such perception is proper when state is conceived as a function assigning a certain value to each cell, but not as this value itself. Thus as in the c-e structures. Another view, when each cell is seen as an automaton changing its state depending on its neighbourhood as an input symbol, will be raised later.

[2] Attention: in this text the indices $x$, $y$ do not mean lower (effect) polynomials of $a$. The notation "$a_{xy}$" (or "$a_{x,y}$") for names (labels) of cells instead of customary "$(x, y)$" in case of 2-dimensional cellular systems is to imitate labelling of entries in matrices (in general "$a_{x,y,\dots,z}$") instead of "$(x, y, \dots, z)$" as well as of points (or vectors) in Cartesian space.

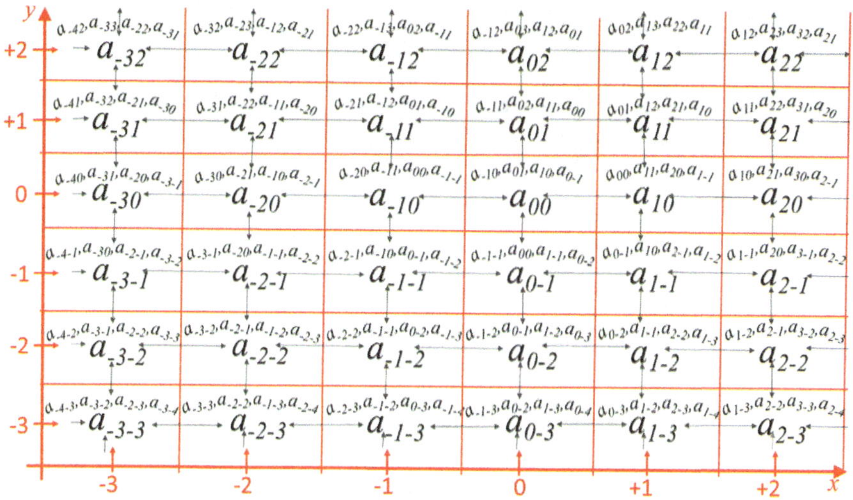


**Fig. 2.1** A fragment of the grid–a plane surface covered by tiles-cells, making a cellular c-e structure called here a "*board*4". The digit "4" indicates rectangular shape of tiles. Each cell is named $a_{xy}$ annotated by its neighbours, i.e. by causes and (the same) effects, written over each node's name in the same order. The indices $xy$ i.e. integer numbers, are not separated by comma–for shortness of notation, hence for the present we write "$a_{xy}$" instead of "$a_{x,y}$". Cell $a_{00}$ is in the coordinate origin

$$C(a_{xy}) = E(a_{xy}) = a_{x-1y} + a_{xy+1} + a_{x+1y} + a_{xy-1}$$

and sign "+" is replaced with comma (",") in Fig. 2.1, to stress that $a_{xy}$ has the set $\{a_{x-1y}, a_{xy+1}, a_{x+1y}, a_{xy-1}\}$ as its neighbourhood, denoted $nbh(a_{xy})$. Each cell is surrounded by the same number of neighbours: $\forall_{x,y}\ |nbh(a_{xy})| = 4$ ($|\ldots|$ means cardinality). A state $s$ is a subset of the set of all nodes.[3] A cell $a_{xy}$ may be active ("live"), in a state $s$, meaning $a_{xy} \in s$ or passive ("dead"), meaning $a_{xy} \notin s$. The next-state rules are:

- The dead cell in a state $s$ with exactly 3 live neighbours, becomes live in the next-state $t$; otherwise remains dead.
- The live cell in a state $s$ with 2 or 3 live neighbours remains live in the next-state $t$; otherwise becomes dead.

The result of passing from state to state concerns all cells simultaneously.

Semantic rules will be made formal after discussion of the rules of formation.

A node with interconnected neighbouring nodes is called an *atom*, or *atomic c-e structure*, and the central node itself–a *nucleus* of the atom, as depicts Fig. 2.2.[4]

---

[3] So, like in case of elementary c-e structures.

[4] What is called here "atomic c-e structure", in the cellular automata area is the neighbourhood of $a_{xy}$ where the nucleus is treated as its own neighbour. In the cellular version of c-e structures, the cell is not its own neighbour. This is justified by combining cellular c-e structures, the issue considered later.

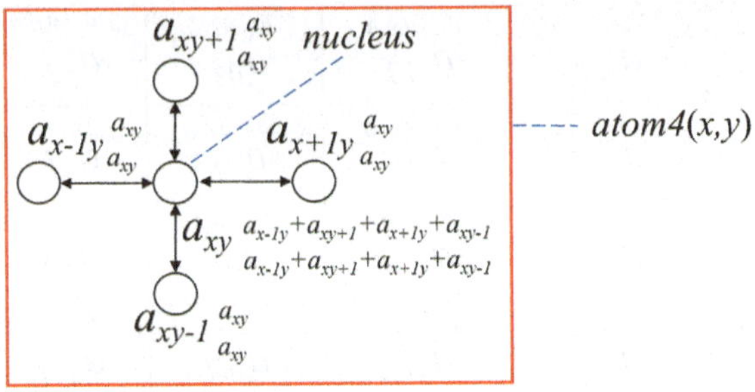


**Fig. 2.2** Atomic c-e structure $atom4(x, y)$ with nucleus $a_{xy}$ and neighbourhood of 4 cells: $nbh(a_{xy}) = \{a_{x-1y}, a_{xy+1}, a_{x+1y}, a_{xy-1}\}$

Without additional annotations, the expression "$atom4(x, y)$" states only the coordinates of nucleus on the 2-dimensional surface and the number of its neighbours, but not their coordinates. A complete meaning of this expression depends on a distribution of neighbours on the surface. In case of the von Neumann type Game of life, the meaning is clarified by tacit (contextual) presumption about location of the neighbours on identical rectangular tiles sharing sides with the nucleus and about covering entire infinite plane surface named "*board*4" by all tiles[5]. In this example– the presumption is assisted by Fig. 2.1. A more explanatory denotation of atomic c-e structure is necessary when a bounded part of a surface hosting a c-e structure is to be extracted. For instance, indicating name of surface hosting a c-e structure, which the atom comes from, like "$\underset{board4}{atom}(x, y)$". This will be needed later. The complete description of $atom4(x, y)$ is the pair $[a_{xy}, \{a_{x-1y}, a_{xy+1}, a_{x+1y}, a_{xy-1}\}]$.

Atomic c-e structures may be added up one to another in accordance with the abovementioned axioms, as shows Fig. 2.3.

The infinite c-e structure, which is a structural base for the Game of life of von Neumann type, is obtained by summation of atomic c-e structures depicted in Fig.2.2, where indices $x$, $y$ range from $-\infty$ to $+\infty$ as shows Fig.2.4. In the algebraic (quasi semiring) notation (Chap. 1) this structural base called "*board*", may be written as the expression:

$$\sum_{-\infty<x,y<+\infty} (a_{xy} \leftrightarrow a_{x-1y}) + (a_{xy} \leftrightarrow a_{xy+1}) + (a_{xy} \leftrightarrow a_{x+1y}) + (a_{xy} \leftrightarrow a_{xy-1})$$

Coming to formalization of the next-state rules for the von Neumann type Game of life, first let us say that a cell is live if the node it represents, holds a token and dead otherwise. The state $s$ of this cellular c-e structure is a subset of the set $\mathbb{X}$ of all

[5] Later on, not only rectangular tiles covering surfaces will be used as cells. Also others, like octagonal, hexagonal, triangular, etc. For the tiling ("tessellation") problems–cf [3].

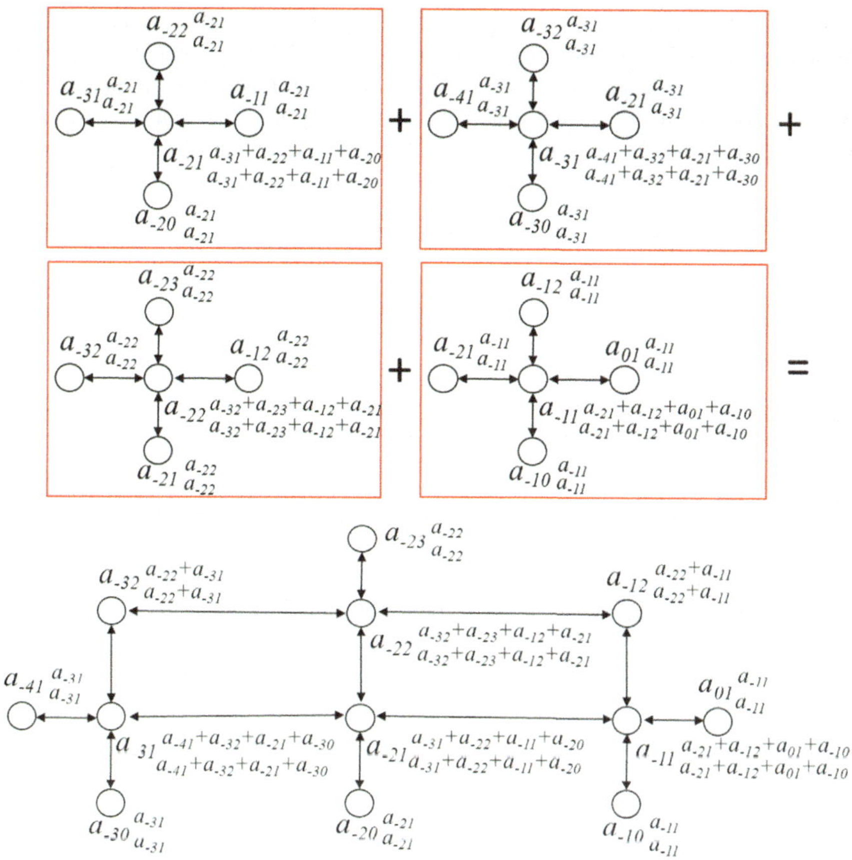


**Fig. 2.3** Summation of four atomic c e structures such as in Fig. 2.2, with nuclei $a_{-21}$, $a_{-31}$, $a_{-22}$, $a_{-11}$

**Fig. 2.4** The infinite c-e structure $\sum_{-\infty<x,y<+\infty} atom\ board4$ $(x, y)$ when imposed on the grid of rectangles, becomes a board for the von Neumann type Game of life. Its fragment is in Fig. 2.1

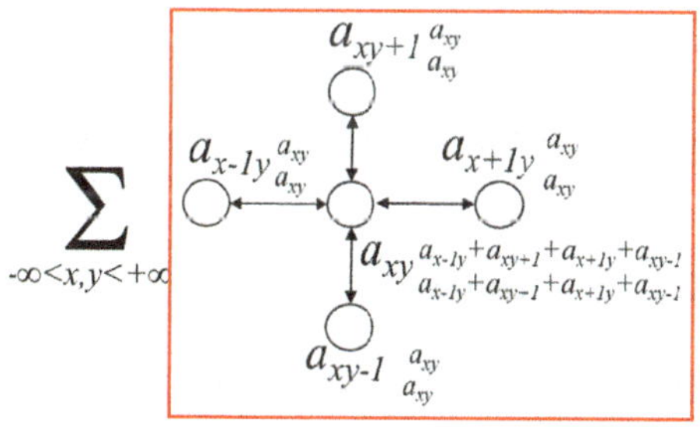

nodes with the meaning: $a_{xy} \in s$ if cell $a_{xy}$ is live and $a_{xy} \notin s$ if $a_{xy}$ is dead.[6] Thus, the intersection of the neighbourhood of $a_{xy}$ with $s$, $nbh(a_{xy}) \cap s$, is the set of live cells, that is, nodes neighbouring $a_{xy}$, active (containing tokens) in the state $s$. Note, that to be consistent with the c-e structures terminology, here the "state" means, what in the cellular automata is called a "configuration". A content of a cell $a_{xy}$, i.e. its active ("live") or passive ("dead") situation in a state $s$, is called a *status* of $a_{xy}$. Let $s$ be a current state and $t$ – the next-state, an effect of one action. Let the binary relation $[\![a_{xy}]\!]$ be defined as:

$$(s, t) \in [\![a_{xy}]\!] \text{ holds if and only if } \underbrace{a_{xy} \in t}_{\text{live } a_{xy}} \text{ and}$$

$$\underbrace{(a_{xy} \notin s \wedge |nbh(a_{xy}) \cap s| = 3)}_{\text{dead } a_{xy} \text{ has 3 live neighbours}} \vee \underbrace{[a_{xy} \in s \wedge (|nbh(a_{xy}) \cap s| = 2 \vee |nbh(a_{xy}) \cap s| = 3)]}_{\text{live } a_{xy} \text{ has 2 or 3 live neighbours}}$$

Then $[\![a_{xy}]\!]$ determines the local (i.e. with no impact on other cells) behaviour of the cell $a_{xy}$ in the state $s$.

The semantics of a cellular c-e structure $U_{vonNeum}$ specifying (or implementing) the Game of life of the von Neumann type is the following binary relation:

$$(s, t) \in [\![U_{vonNeum}]\!] \Leftrightarrow \forall a_{xy} \in car(U_{vonNeum}) : (s, t) \in [\![a_{xy}]\!]$$

where in this case the carrier of $U_{vonNeum}$ coincides with the universe $\mathbb{X}$ of all nodes (cells): $car(U_{vonNeum}) = \mathbb{X}$ (see Chap. 1).

#### 2.1.1.2 The Game of Moore Type

This type of game differs from the former only by the number of neighbours of cells: $\forall_{x,y}\ |nbh(a_{xy})| = 8$. So, the atomic c-e structure looks now as in Fig. 2.5 and the next-state rule is as in case of the von Neumann type. The semantics of a cellular c-e structure $U_{Moore}$ specifying (or implementing) the Game of life of the Moore type is the binary relation $[\![U_{Moore}]\!]$ defined by the same formula as for $[\![U_{vonNeum}]\!]$.

#### 2.1.1.3 Remarks

If the cells are to be imagined as some geometric shapes, polygonal usually, then their spatial distribution may enjoy features such as sharing (adjoin) sides or/and vertices

---

[6] Obviously, the state may be treated as a function $s : \mathbb{X} \rightarrow \{\mathbf{0}, \mathbf{1}\}$ if the values $\mathbf{0}, \mathbf{1}$ reflect somehow conditions $a \notin s$, $a \in s$. For instance, when we fix that $s(a) = \mathbf{0}$ means $a \notin s$ and $s(a) = \mathbf{1}$ means $a \in s$. Such treatment is used later in description of state for general cellular c-e structures, where the members of range of function $s$ are called states–in accordance with terminology of cellular automata.

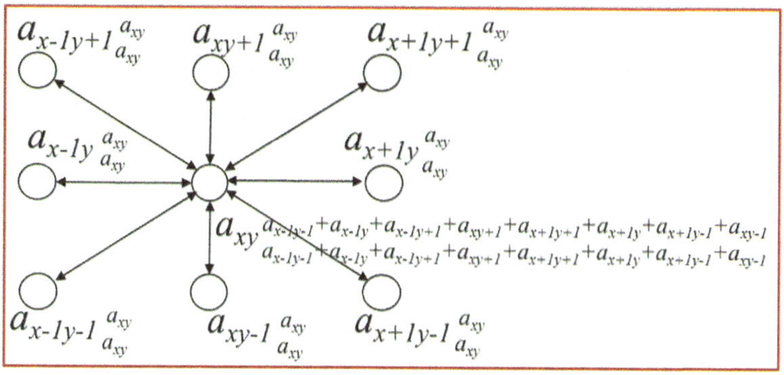

**Fig. 2.5** Atomic c-e structure *atom*8$(x, y)$ with nucleus $a_{xy}$ and neighbourhood of 8 cells: $nbh(a_{xy}) = \{a_{x-1y-1}, a_{x-1y}, a_{x-1y+1}, a_{xy+1}, a_{x+1y+1}, a_{x+1y}, a_{x+1y-1}, a_{xy-1}\}$

with neighbouring cells. Such features are exploited later when cut-out a subset of all cells covering plane surface is an intermediate structure for placing cellular c-e structures on curved surfaces like cylinders and toruses. In Fig. 2.1 each rectangular cell ("tile") shares sides with all its neighbours in case of the von Neumann type game. In the Moore type, each rectangular cell shares sides only with four neighbours and the vertices also only with four. In Fig. 2.6, geometrically related with the Moore type game, each octagonal cell shares sides only with four neighbours, whereas is disjoint with four remaining. In this case, the atomic c-e structure is

$$[a_{xy}, \{a_{x-1y-1}, a_{x-1y}, a_{x-1y+1}, a_{xy+1}, a_{x+1y+1}, a_{x+1y}, a_{x+1y-1}, a_{xy-1}\}].$$

Note however that resigning from identical shape of cells, so letting to cover plane by different polygons-cells, the sharing of sides of each cell with neighbouring cells may be ensured, as shows Fig. 2.7, but this might make loss of basic property: equinumerous neighbourhoods and even loss of some nodes! The association of cellular c-e structures with geometric shape of cells and features like sharing sides or surface area of atoms, though expedient to assimilate main ideas of the theory (and suitable for some applications), plays the illustrative role only. In the formal terms, concepts like neighbourhood, atomic c-e structure, etc. are graphs. Nonetheless, attention will be given to the 1-1 correspondence between sides of each cell-tile and of its neighbours. So, to the adherence of cell's sides to sides of its neighbours. In other words–to sharing sides of each tile with surrounding neighbours. The reason of such assumption is rooted in using addition in algebra of c-e structures to make curved bases for cellular c-e structures. The problem of feasibility to achieve this will be touched in Chap. 3. To cover a surface with one kind of polygonal cells to obtain a cellular c-e structure not always is possible, which illustrates Fig. 2.7.

Another remark concerns naming of cells. If we imitate notation customary for matrices, the coordinate axes on the plane surface, correspond to rows and columns. The entries (that assume values of states) are located at points of integer coordinates.

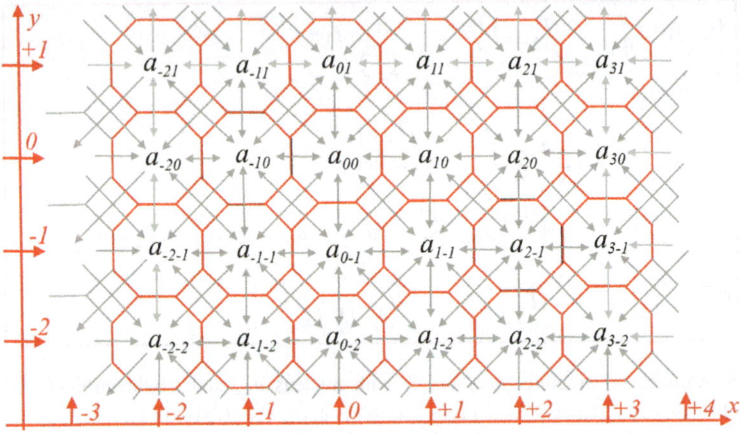


**Fig. 2.6** A fragment of the infinite planar surface *board*8, tiled with octagons and squares. The octagons only, not squares, are the cells of this cellular c-e structure. Each octagonal cell (node) is linked to its neighbours, i.e. causes and (the same) effects, those depicted in Fig. 2.5. All atomic c-e structures are equivalent relative-to the shift transformation

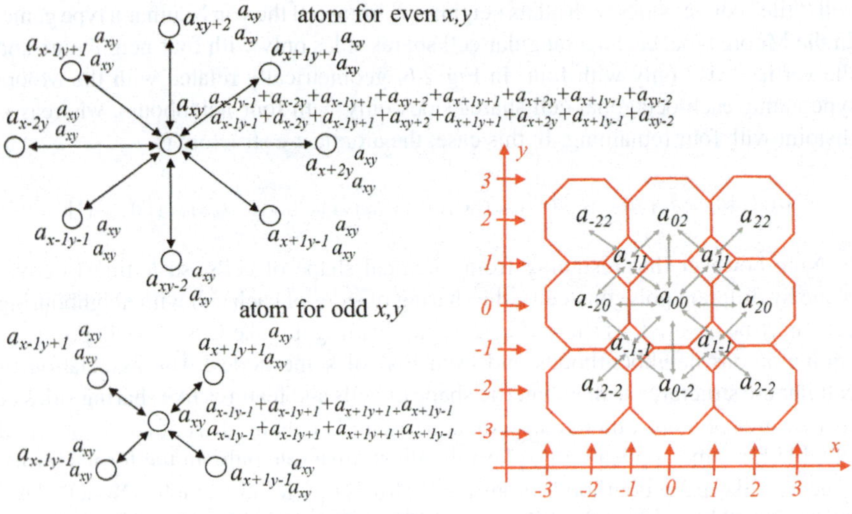


**Fig. 2.7** Atoms with nuclei on octogonal cells of even coordinates, so having 8 neighbours, as well atoms with nuclei on square cells of odd coordinates, so having 4 neighbours. Due to different than in Fig. 2.6 association of the coordinates to cells, the cells share sides with their neighbours: the infinite plane is tessellated by octagons and squares. However such geometric arrangement does not lead to <u>cellular</u> c-e structures, but only to ordinary ones. A red fragment of such tessellation shows one atom with 8 neighbours and four atoms with 4 neighbours each

The entries of matrix $a$, thus cells, are denoted by $a_{xy}$ instead of simply $(x, y)$, because entry $b_{xy}$ of matrix $b$ located at the same point as $a_{xy}$, may be surrounded by other neighbours. This happens when summing up different ordinary c-e structures, so as to get a desirable cellular c-e structure. For instance, a piece cut out from a cellular c-e structure and stretched on Cartesian plane, has different matrix, in which cells internal to this piece have the same neighbours as in the original cellular c-e structure, but fewer than cells on the boundary. This case will be dealt with later, when the flat cellular c-e structures are twisted up into cylindric and toroidal surfaces. Notice also that in some tessellations of the plane, some points $(x, y)$ are not hosting cells–see Figs. 2.7 and 2.9.

**End of example "Game of life"**.

In the "Game of life" cellular c-e structure, a behaviour of any cell (node) $a_{xy}$, i.e. change (or not) its passive/active status into opposite, depends solely on its content and on number of its active neighbours, disregarding their position in the neighbourhood. That is, on a cardinality of subsets of $nbh(a_{xy})$ of active neighbours in a state $s$: 3 if $a_{xy} \notin s$ and 2 or 3 if $a_{xy} \in s$. Obviously, in the von Neumann case, there are $\binom{4}{3} = \frac{4!}{3!(4-3)!} = 4$ three-membered subsets of the set $nbh(a_{xy})$ and $\binom{4}{2} = \frac{4!}{2!(4-2)!} = 6$ two-membered subsets of the set $nbh(a_{xy})$, whereas, in the Moore case, these numbers are $\binom{8}{3} = 56$ and $\binom{8}{2} = 28$ respectively. This prompts some generalizations of this example cellular c-e structure.

## 2.2 A Slight Generalization

Consider cellular c-e structures with cells identified by coordinates $x, y$ and with ***false*** (empty cell) or ***true*** (token in) as admitted status, where the next-state rule depends only upon the cell's status and upon a number of active neighbours.

Let in a cellular c-e structure $U$ with carrier $\mathbb{X}$, where $\forall_{x,y}\ |nbh(a_{xy})| = n$ (size of neighbourhood), a set of numbers $B \subseteq \{1, 2, \ldots, n\}$ points out how many active neighbours of a cell ensure "Birth" and a set of numbers $D \subseteq \{0, 1, 2, \ldots, n\}$ – how many active neighbours ensure "Duration" (survival). The next-state rule for the cell $a_{xy}$ is the binary relation $[\![a_{xy}]\!]$: $(s, t) \in [\![a_{xy}]\!]$ holds if and only if:

$$a_{xy} \in t \wedge [(a_{xy} \notin s \wedge |nbh(a_{xy}) \cap s| \in B) \vee (a_{xy} \in s \wedge |nbh(a_{xy}) \cap s| \in D)] \quad (BD)$$

Thus the semantics $[\![U]\!]$ of $U$ is: $\forall a_{xy} \in car(U) : (s, t) \in [\![a_{xy}]\!]$. If $U$ is the Game of life then $B = \{3\}$, $D = \{2, 3\}$.

Let, for example $\forall_{x,y}\ |nbh(a_{xy})| = 6$. Such atomic c-e structure $U$ looks like in Fig. 2.8. Let us also assume that the next-state rule depends only upon the cell's state and upon a number of active neighbours.

The fragment of infinite planar base for cellular c-e structure being the sum of atoms depicted in Fig. 2.8, where indices $x, y$ range from $-\infty$ to $+\infty$ is the c-e structure in Fig. 2.9. The natural pictorial shape of each cell in this case is a hexagon. Since the Euclidean distance between a cell and its neighbours is immaterial, the scaling of axes $x, y$ of Cartesian coordinates may differ in pictorial presentation of

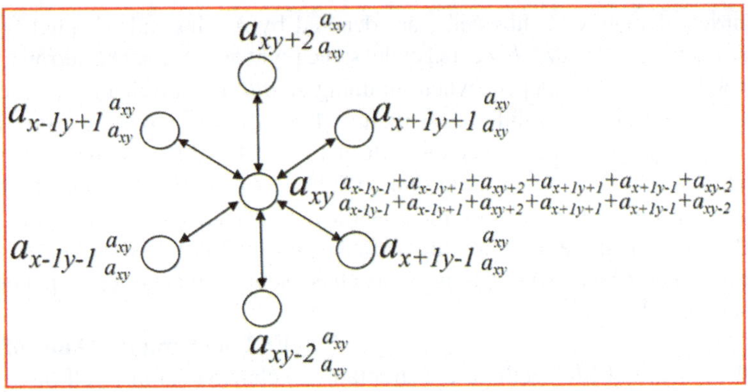


**Fig. 2.8** Atomic c-e structure *atom*6$(x, y)$ with nucleus $a_{xy}$ and neighbourhood of 6 cells: $nbh(a_{xy}) = \{a_{x-1y-1}, a_{x-1y+1}, a_{xy+2}, a_{x+1y+1}, a_{x+1y-1}, a_{xy-2}\}$. This atom, if taken from the planar surface *board*6, will also be written as $\underset{board6}{\underline{atom}}(x, y)$

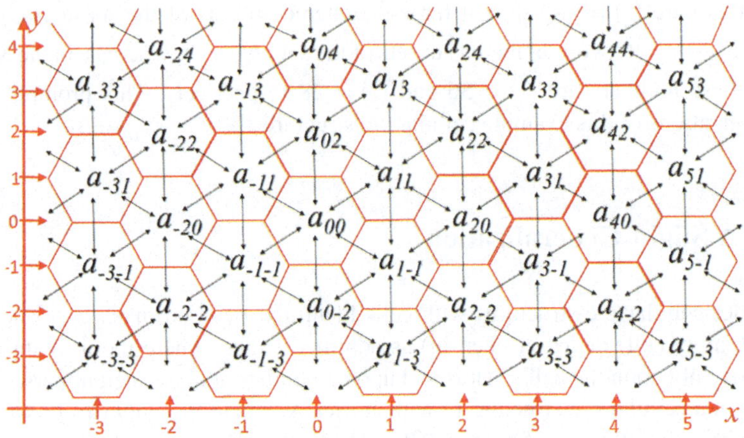


**Fig. 2.9** A fragment of the infinite planar surface "*board*6". The digit "6" indicates hexagonal shape of tiles-cells. Each cell $a_{xy}$ is linked to its neighbours, i.e. causes and (the same) effects, as in Fig. 2.8. Note that in such arrangement, only cells with the same parity of $x$ and $y$ exist. All atomic c-e structures are equivalent relative-to the shift transformation (likewise in cellular automata)

planar bases for cellular c-e structures, like in Figs. 2.1, 2.6, 2.7 or Fig. 2.9.[7] Various cellular c-e structures may be specified by fixing sets $B$ and $D$ of natural numbers. Examples of evolution of cellular c-e structure placed on hexagonal cells and with chosen sets $B$, $D$, are depicted in Figs. 2.10 and 2.11.

An important remark on geometric arrangement of cells on a surface should be made, if tiling [3] by cells of <u>identical</u> shape is advisable. That is, if every cell is to share sides with its neighbours, as in Fig. 2.9, which not always is possible. For instance when cells of the same shape have 8 neighbours–see Figs. 2.5 and 2.6–

[7] So, a square may be seen as a rectangle, a circle as an ellipse, etc. Such shapes are affine equivalent.

thus, should look like octagons. In this case, parity of their coordinates in rows (along $x$ axis) and columns (along $y$ axis) should be regarded. Such case concerns also hexagons, as shows Fig. 2.9, where every second row and column of cells have the same parity of coordinates, but the parity of succeeding is alternating. In such arrangement, in some points $(x, y)$ no cells exist, like in Fig. 2.9. Such arrangement will be applied for examples of transformation of planar surface into cylinders and toruses, as shown later. Algebraically, the base in Fig. 2.9 may be described as the sum of atoms in Fig. 2.8, as follows:

$$\underbrace{\sum_{-\infty<x<+\infty} \sum_{-\infty<y<+\infty} \underset{board6}{atom}(x, y)}_{\text{for even } x \text{ and } y} + \underbrace{\sum_{-\infty<x<+\infty} \sum_{-\infty<y<+\infty} \underset{board6}{atom}(x, y)}_{\text{for odd } x \text{ and } y}$$

where $\underset{board6}{atom}(x, y) = [a_{xy}, \{a_{x-1y-1}, a_{x-1y+1}, a_{xy+2}, a_{x+1y+1}, a_{x+1y-1}, a_{xy-2}\}]$.

### 2.2.1 *Example (Explosion)*

Evolution of cellular c-e structure on the base in Fig. 2.9 (thus $|nbh(a_{xy})| = 6$) with the next-state rule $(BD)$, where $B = \{1, 2\}$, $D = \{0, 1, 2, 3, 4, 5, 6\}$ and with initial state $s_0$ such that $a_{00} \in s_0$ and $a_{xy} \notin s_0$ for $x \neq 0$, $y \neq 0$, is shown in Fig. 2.10.

Here too, change of passive/active status of a cell, depends exclusively on its status and on number of its active neighbours, not on their position in the neighbourhood. Each cell has $\binom{6}{1} = 6$ one-membered and $\binom{6}{2} = 15$ two-membered subsets of $nbh(a_{xy})$. If in the state $s$, a cell $a_{xy}$ is passive and has 1 or 2 active neighbours, then becomes active in the next-state $t$, otherwise remains passive. If $a_{xy}$ is active in the state $s$ and has no more than 6 active neighbours, then remains active, otherwise becomes passive (which never happens in this example).

**End of example.**

### 2.2.2 *Example (Cog-Wheel)*

Changing parameter $B$ in the firing rule $(BD)$ so that $B = \{2\}$, leaving $D$ without change, and changing the initial state $s_{init}$ so that now the cells $a_{00}$, $a_{-1-1}$, $a_{-11}$, $a_{02}$, $a_{11}$, $a_{1-1}$, $a_{0-2}$ are active (i.e. belong to $s_{init}$) but the remaining are passive, the explosion progress shown in Fig. 2.10 halts at cells $a_{-20}$, $a_{-13}$, $a_{13}$, $a_{20}$, $a_{1-3}$, $a_{-1-3}$, as shows Fig. 2.11.

**End of example.**

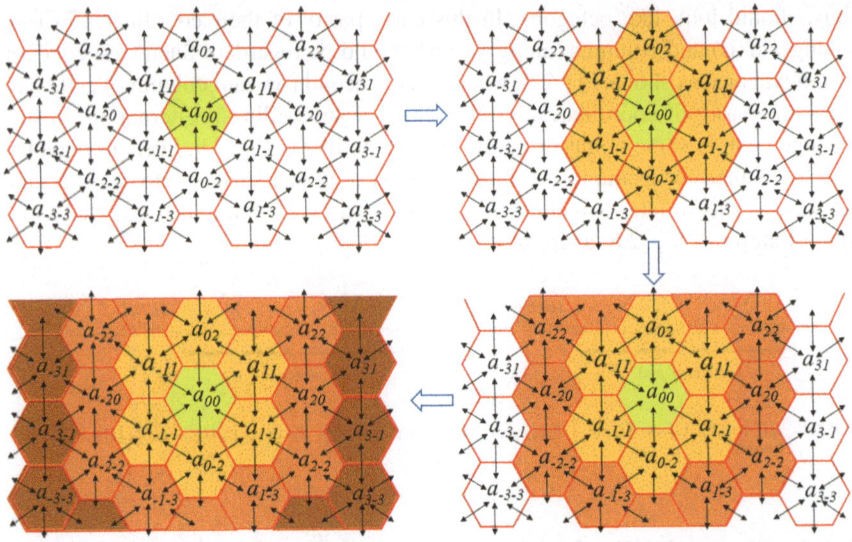

**Fig. 2.10** Four successive stages of evolution of cellular c-e structure based on Fig. 2.9, starting with the single active cell $a_{00}$ and semantics $(BD)$ with $B = \{1, 2\}$, $D = \{0, 1, 2, 3, 4, 5, 6\}$. The consecutive (yellow→orange→red→brown→…) area blocks of active cells expand ex-centrally in the isotropic way, in accordance with the next-state rule $(BD)$. No terminal state

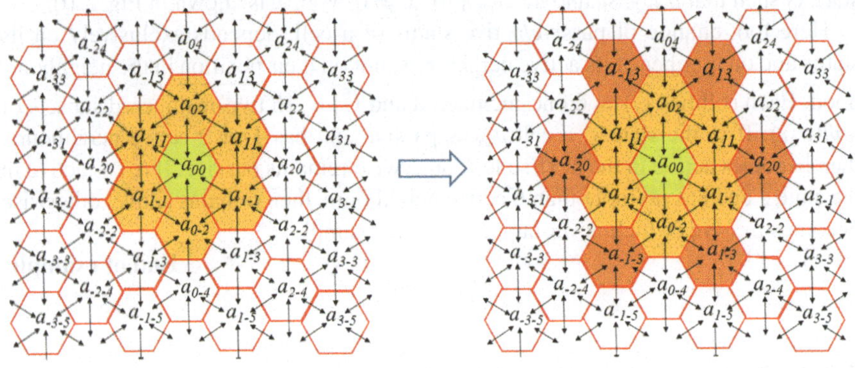

**Fig. 2.11** Evolution of cellular c-e structure on the base in Fig. 2.9 with the next-state rule $(BD)$, where $B = \{2\}$, $D = \{0, 1, 2, 3, 4, 5, 6\}$ and with initial state $s_{init} = \{a_{00}, a_{-1-1}, a_{-11}, a_{02}, a_{11}, a_{1-1}, a_{0-2}\}$ (yellow and orange cells), and terminal state (yellow, orange and red cells), reminding a cog-wheel

## 2.3 Further Generalization

For a cell $a_{xy}$, the new parameters denoted by $B1$ and $D1$ will be now collections (families) of subsets of $nbh(a_{xy})$, that is $B1 \subseteq 2^{nbh(a_{xy})}$ and $D1 \subseteq 2^{nbh(a_{xy})}$. It should be noticed that for the empty set {}, the family $2^{\{\}}$ is non-empty, because its single element is {}: $2^{\{\}} = \{\{\}\} \neq \{\}$. Instead of the foregoing next-state rule $(BD)$ for a cell $a_{xy}$, let us consider the following one:

$(s, t) \in [\![a_{xy}]\!]$ holds if and only if:

$$a_{xy} \in t \wedge [(a_{xy} \notin s \wedge nbh(a_{xy}) \cap s \in B1) \vee (a_{xy} \in s \wedge nbh(a_{xy}) \cap s \in D1)] \quad (B1D1)$$

As before, the semantics of a cellular c-e structure $U$ is:

$$\forall a_{xy} \in car(U) : (s, t) \in [\![a_{xy}]\!].$$

The intended[8] role of family $B1$ is that it makes the passive cell $a_{xy}$ in a state $s$ active in the next-state $t$, if the set $nbh(a_{xy}) \cap s$ of active in $s$ cells neighbouring $a_{xy}$ is in $B1$; otherwise $a_{xy}$ remains passive. Similarly, the family $D1$ makes the active cell $a_{xy}$ in the state $s$ to remain active in the next-state $t$, if the set of active cells neighbouring $a_{xy}$ is in $D1$; otherwise $a_{xy}$ becomes passive. So, behaviour of a cell is determined not only by a number of its active neighbours, but by a placement (position) of active cells in its neighbourhood, specified by means of the families $B1$ and $D1$. One may say that semantics $(B1D1)$ is more selective than $(BD)$, which is more neglectful. Thus, $(B1D1)$ is more general than $(BD)$ indeed: for parameters $B$, $D$, it is always possible to choose $B1$, $D1$ so that every task behaving under the next-state rule $(BD)$ would behave identically under $(B1D1)$. The next-state rule[9] selection for specific tasks amounts to determine sets $B$, $D$ or families $B1$, $D1$.

It is not difficult to count how many next-state local rules exist, if $|nbh(a_{xy})| = n$ and each cell assumes two values, ***true*** (active), or ***false*** (passive). The local rule determining behaviour of the cell $a_{xy}$ is the binary relation $[\![a_{xy}]\!]$ between state $s$ and the next-state $t$ defined by $(s, t) \in [\![a_{xy}]\!]$ and specified by the $(B1D1)$ rule. There are $2^n$ subsets of $nbh(a_{xy})$ and $a_{xy}$ is either ***false*** in the state $s$ ($a_{xy} \notin s$) or ***true*** ($a_{xy} \in s$). Here, the cell $a_{xy}$, though takes part in transition from $s$ to $t$, has not been included into $nbh(a_{xy})$. Counting the cell $a_{xy}$ together with its neighbours, there are $2^{n+1}$ subsets of the set of all cells involved, each cell passive or active in $s$. Considering all cases involved in the rule $(B1D1)$, yields $2^{2^{n+1}}$ relations $[\![a_{xy}]\!]$. Note that there are $2^{2^{n+1}}$ Boolean functions of $n + 1$ arguments and that each such function corresponds to one relation $[\![a_{xy}]\!]$. Computing logical functions by cause-effect structures is described in [1].

[8] By designer of a cellular c-e structure for a task.

[9] Parameters $B1$, $D1$ play part of the local rule $f$ in literature on cellular automata.

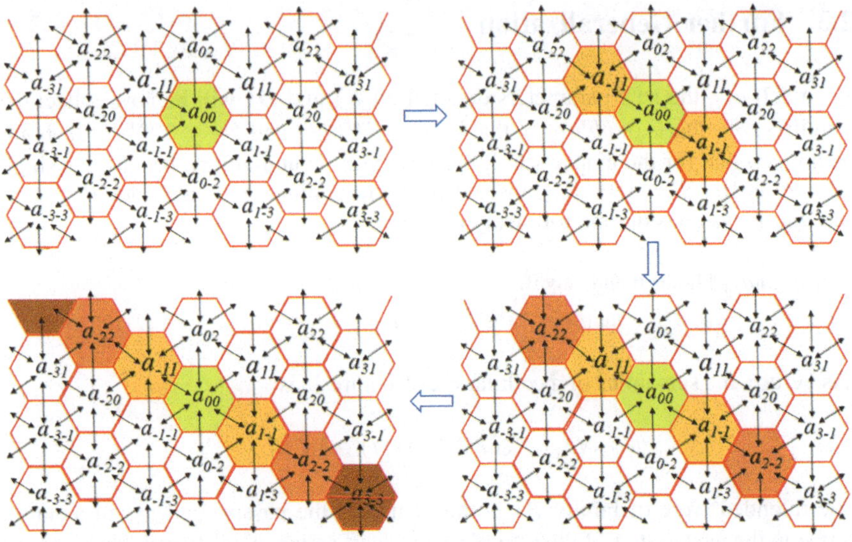


**Fig. 2.12** Four successive stages of evolution of cellular c-e structure based on Fig. 2.9, starting with the single active cell $a_{00}$ and semantics $(B1D1)$ with $B1 = \{\{a_{x+1y-1}\}, \{a_{x-1y+1}\}\}$, $D1 = \{\{\}, \{a_{x+1y-1}\}, \{a_{x-1y+1}\}\}$. The consecutive (yellow→orange→red→brown→ ...) darkening coloured cells pretend two light rays, propagating from the source $a_{00}$ in two reverse directions, in accordance with the next-state rule $(B1D1)$. Indeed, if the cells were very small, almost as points, their coloured sequences would look like the two light rays trajectories

So, in examples of cellular c-e structures so far shown, the number of local next-state rules was as follows[10]: for $|nbh(a_{xy})| = 4, 6, 8$ the number of local next-state rules are respectively:

$$2^{2^5} = 2^{32} = 4294967296,\ 2^{2^7} = 2^{128},\ 2^{2^9} = 2^{512}.$$

Note that here neighbours of cells are not ordered: any neighbourhood is a set, not a vector, like in the theory of cellular automata.

### 2.3.1 *Example (Two Light Rays)*

Evolution of cellular c-e structure based on Fig. 2.9 with the next-state rule $(B1D1)$, where $B1 = \{\{a_{x+1y-1}\}, \{a_{x-1y+1}\}\}$, $D1 = \{\{\}, \{a_{x+1y-1}\}, \{a_{x-1y+1}\}\}$ and with initial state $s_0$ such that $a_{00} \in s_0$ and $a_{xy} \notin s_0$ for $x \neq 0$, $y \neq 0$, is shown in Fig. 2.12.

**End of example.**

[10] The same rule with various parameters $B$, $D$ governs a class of tasks, such as "Game of life", "explosion", "cog-wheel", or with parameters $B1$, $D1$ like further tasks "light rays". A concrete task is specified by fixing cells' neighbourhoods (shape of cells perhaps) and initial state.

### 2.3.1.1 Remark

The base for exemplary cellular c-e structures considered so far was perceived as Cartesian plane surface with cells treated as points surrounded by neighbours. In such perception, the parameters $B1$, $D1$ make the active cells overlie the straight line $\{\langle x, y\rangle : y = -x\}$ in Fig. 2.12. The modified parameters as follows: $B1 = \{\{a_{x+1y+1}\}, \{a_{x-1y-1}\}\}$, $D1 = \{\{\}, \{a_{x+1y+1}\}, \{a_{x-1y-1}\}\}$ would make the active cells overlie the orthogonal straight line $\{\langle x, y\rangle : y = x\}$. The parameters $B1$, $D1$ and $B$, $D$ in the next-state rules $(B1D1)$ and $(BD)$, with various initial states, create various geometric objects as sets of active cells. For instance, the set of all cells in case of Fig. 2.10, whereas in case of Fig. 2.11–the set of active cells (a "cog-wheel") that overlie the following set of points $\langle x, y\rangle$:

$$\underbrace{\{\langle x, y\rangle : |x| \leq 2 \wedge |y| \leq 3 \wedge |x| \neq |y|\}}_{\text{boundary of the "cog-wheel"}} \cup \underbrace{\{\langle x, y\rangle : |x| \leq 1 \wedge |y| \leq 2\}}_{\text{inside of the "cog-wheel"}}$$

here $|\ldots|$ denotes the absolute value, not cardinality.

In the next chapter, certain distortions of the Cartesian 2-dimensional surface covered by cells will be made, using addition of c-e structures, the operation axiomatically characterized by Eqs. (2.1) at the beginning of this chapter.

## References


1. Czaja, L.: Computing in Cause-Effect, Structures. Lecture Notes in Networks and Systems, vol. 331. Springer (2022)
2. Gardner, M.: The fantastic combination of John Conway's new solitaire game "life." Sci. Am. **223**, 120–123 (1970)
3. Grunbaum, B., Shepard, G.C.: Tilings and Patterns. Dover Publications Inc., Mineola New York (2016)
4. Moore, E.F.: Machine models of self-reproduction. In: Proceedings of Symposia in Applied Mathematics XIV: "Mathematical Problems in Biological Sciences". AMS, pp. 17–33
5. Von Neumann, J.: Theory of Self-Reproducing Automata. Published by University of Illinois Press, Urbana, IL (1966). (Edited and completed by Arthur W. Burks)
6. Zaitsev, D.: Simulating cellular automata by infinite petri nets. J. Cell. Autom. (2018)

# Chapter 3
# Structures on Non-planar Surfaces

## 3.1 Preliminary Remarks

1. Contrary to the ordinary cause-effect structures outlined in Chap. 1, summation of the cellular c-e structures not always is meaningful. The result not always is a cellular c-e structure, neither structurally nor semantically. So, one cannot say of cellular c-e structures as of an algebra with the add operation (+) obeying axioms (2.1) in Chap. 2, as in case of the standard c-e structures. Sometimes addition yields unequal number of neighbours of cells, and the next-state rules are ordinarily specific, autonomous for different addends. Nonetheless, summing up c-e structures, may result in a correct cellular c-e structure. Such case will be exploited in constructing cellular c-e structures outspread on some non-planar surfaces like cylinders and toruses.
2. Geometric construction of a cellular c-e structure on a curved surface will take place by extraction (cut out) of a segment of Cartesian plane covered by cells and complementing it–by means of summation–with suitably chosen auxiliary c-e structures, to obtain a correct cellular c-e structure. The segment will be an array of cells extracted along one or both axes, i.e. a finite or infinite rectangle-like shape. Complementation of its sides with atomic c-e structures taken from opposite sides, makes these sides meet, giving a correct cellular c-e structure. This simulates placing ("overwrap") the array on a closed rounded surface, cylindric and toroidal.
3. Extracting a piece of the full cellular c-e structure outspread on Cartesian plane shown e.g. in Figs. 2.1, 2.9 (Chap. 2), is accomplished by selecting a set of atomic c-e structures and making their sum.[1] For instance,

[1] Contrarily to summation in the quasi semiring, where apart from the addition, multiplication of c-e structures is possible, here, in the commutative and idempotent monoid, the summation never creates firing components, apart from those present in the constituents.

 
L. Czaja, *Cellular Cause-Effect Structures*, Studies in Systems, Decision and Control 526,
https://doi.org/10.1007/978-3-031-55590-9_3

$$rectangle \stackrel{\text{def}}{=} \sum_{x1 \le x \le x2} \sum_{y1 \le y \le y2} \underset{board4}{atom}(x, y)$$

specifies a c-e structure of rectangular shape with vertices $(x1, y1)$, $(x2, y1)$, $(x1, y2)$, $(x2, y2)$–a piece cut out from the surface *board*4 in Fig. 2.1. Here, the atomic c-e structure $\underset{board4}{atom}(x, y)$ inside the rectangle is a cell surrounded by 4 neighbours, whereas on its sides it is a cell surrounded by 3 or 2 neighbours (2 in the corners)–cut off please, a rectangle from the *board*4 in Fig. 2.1. Selection of the atomic structures from the *board*4, amounts to specifying a carrier of the c-e structure outspread on this rectangle:

$$car(rectangle) = \{a_{xy} : x1 \le x \le x2,\ y1 \le y \le y2\}$$

The cells $a_{xy}$ are nuclei of the selected atoms. The underline "*board*4" of $atom(x, y)$ indicates which c-e structure the atom is taken from. Indicating such c-e structure will be necessary for combining *rectangle* with atoms taken from its opposite sides, to implement its bending.

4. Remember that due to the idempotency of addition, cells identically named in the addends (i.e. of identical coordinates) become one cell of such name in the result–see Fig. 2.3.
5. If cells are to share sides with neighbours, then specifying successor/predecessor relationship of cells along coordinates depends on the cells' geometric shape. So–when the number of each cell neighbours equals the number of its sides. This happens in case of atoms shown in Fig. 2.2 (with rectangular cells) and Fig. 2.8 (with hexagonal) but not in Fig. 2.5. For rectangular cells like in Fig. 2.1, their successor/predecessor relationship along axes corresponds directly to succession of their coordinates. But this is not so with hexagonal, as shows Fig. 2.9, where coordinates of any cell and its successor along one axis differ by 2. This substantiates summation of such atoms along axes separately for even and odd coordinates.
6. In this chapter, the cellular c-e structures of hexagon-shaped cells sharing sides with their neighbours will be considered. Obviously, there are only three types of regular[2] polygons which can tile (by means of shifts only) the plane surface with one of their type and size. These are triangular, rectangular and hexagonal shaped. So, only such cells can share sides with their neighbours. For demonstration of wrap-around flat surfaces onto cylinders and toruses, the cellular c-e structures

[2] The rectangle and square are indistinguishable here; the rectangle is treated as regular. This is in accord with admissible different scaling of the $x$ and $y$ coordinates.

with hexagonal cells will be taken.[3] This seems more instructive when combining c-e structures and geometrically more attractive[4] than in case of rectangular cells.

## 3.2 Bending the Planar Bases

As a base (ground) for pictorial presentation of cellular c-e structures considered so far, have been taken the infinite planar surface with Cartesian coordinates but with ignored Euclidean magnitudes like distance, angle, area, etc. Hence allowable different scaling of the coordinates. Thus possible varied shapes of affinely equivalent polygons that envision cells, e.g. rectangles or hexagons do not have to be equilateral (regular). Algebraically, a cellular c-e structure on a base was the result of infinite summation of atoms shown in Figs. 2.2, 2.5, 2.8 in Chap. 2. Now, let us consider bases bounded along the horizontal direction as in Fig. 3.1 and in both directions–horizontal and vertical. It should be remembered that cause-effect structures have been defined as infinite objects, each based on the infinite space $\mathbb{X}$ of nodes. Some nodes of a c-e structure $U$ are in the carrier $car(U)$, those at the outside, "isolated", belong to $U$ but do not impact behaviour of $U$. Therefore, construction of a bounded c-e structure does not require extra concepts, but only choosing its carrier needed for concrete tasks. Figuratively speaking–to extract from the infinite planar base a piece, being an "array" (or "belt"), infinite or finite.

Nodes-cells on boundary of such array, have less neighbours than those inside. Thus the array, though a c-e structure, is not a cellular c-e structure. This discrepancy disappears if the array's two sides meet, in the sense that cells on one side receive additional neighbours taken from opposite side. Geometrically, such transformation makes a cylinder from this planar array–as if it were "wrapped-around" one axis. Algebraically this is obtained by summing up the planar array with auxiliary c-e structure made of cells from the array's sides. Let us illustrate this by examples with hexagon shape of cells tiling on several curved surfaces.[5] In Fig. 3.1, is a c-e structure called *belt*, with 4 cells in $y$-direction and of infinitely many cells in $x$-direction. So,

[3] It is known that cylinders and toruses can be tessellated by identical hexagons, but spheres cannot. This follows from the classic Euler characteristic formula and more involved algebraic-topological considerations [1, 4]. So, no sphere can be a base for cellular c-e structure with hexagonal shaped cells. Obviously, the sphere can be tessellated by hexagons and pentagons (like some soccer balls or the dome over the control tower of Warsaw's airport). Such cover of spheres violates requirement on equal number of neighbours of every cell, which is inadmissible for cellular c-e structure, but admissible for the ordinary ones.

[4] Also the bees think so. It might also be remarked that spherical cup being the mirror of the Webb telescope is tessellated by regular hexagons.

[5] The geometric shape of cells is immaterial for formal essence of cellular c-e structures but, perhaps, useful for intuitive comprehension of their peculiarity. Sometimes also beneficial owing to similarity of cells' shape to their archetypes from nature, like in case of honeycombed. That is why it is expedient that the cell's shape corresponds to the number of its neighbours, e.g. the rectangle in case of 4 neighbours, the hexagon in case of 6 neighbours. Note that this is not so with the Moore type of Game of life: not all neighbours of a cell are cells with adjacent sides with it.

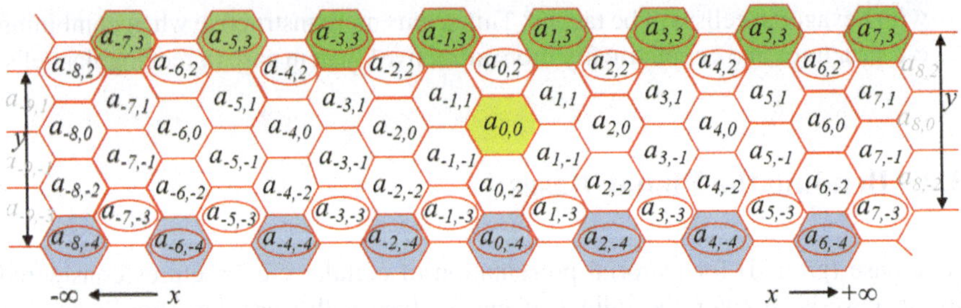


**Fig. 3.1** C-e structure *belt* of 4 cell in width and infinite in length, extracted from *board*6, covering infinite area of cells bounded along axis $x$ inbetween sequences of blue and green cells $[a_{x,-4}]_{(-\infty<x<+\infty \text{ for even } x)}$ and $[a_{x,3}]_{(-\infty<x<+\infty \text{ for odd } x)}$. Ordinary (non-cellular) c-e structure. The yellow cell $a_{0,0}$ is in the coordinate origin. Extreme cells of the columns, i.e. $a_{x,-4}$, $a_{x,2}$ (even $x$) and $a_{x,-3}$, $a_{x,3}$ (odd $x$), labelled inside the ellipses, will be neighbours, when the belt becomes "wrapped-around" the axis $x$ into the infinite cylinder

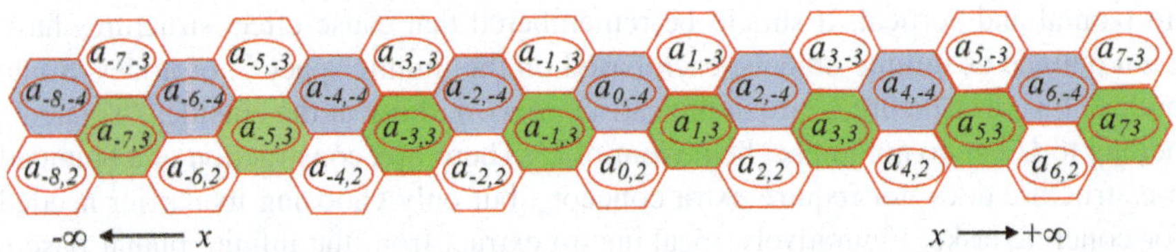


**Fig. 3.2** The auxiliary c-e structure *yjoint* composed of cells taken from horizontal boundaries of *belt*. The cells belonging to *yjoint* have 5 and 3 neighbours. When added to *belt* (that in Fig. 3.1), the image of infinite cylinder arises, flatten (outspread) in Fig. 3.3 and stereo in Fig. 3.4

the set of cells, extracted from the infinite area *board*6 in Fig. 2.9, Chap. 2. This is the infinite c-e structure (not yet cellular one) bounded by sequences of (blue) cells $[a_{x,-4}]$ with even $x$ and (green) $[a_{x,3}]$ with odd $x$ (for better readability, the coordinates $x$ and $y$ in subscripts of "$a$" are separated by comma). Since connection between cells is determined by adherence (i.e. sharing) of their sides, the bidirection arrows, such as in Fig. 2.9, as redundant, are omitted. The atomic c-e structures with nuclei $a_{x,y}$ where $-\infty < x < \infty$, $y = -2, -1, 0, 1$ inside the *belt*, have 6 neighbours, whereas those outside, with $-\infty < x < \infty$, $y = -4, -3, 2, 3$, have them 3 or 5, which is the consequence of ordinary, commutative and idempotent summation of c-e structures.

The bounded c-e structure *belt* in Fig. 3.1, is the sum of its atomic cells:

$$belt = \underbrace{\sum_{-\infty<x<+\infty} \; \sum_{-4\leq y\leq +2}}_{\text{for even } x,y} atom_{\text{even in } belt}(x, y) + \underbrace{\sum_{-\infty<x<+\infty} \; \sum_{-3\leq y\leq +3}}_{\text{for odd } x,y} atom_{\text{odd in } belt}(x, y)$$

$\underset{\text{even in } belt}{atom}(x, y)$ is a cell $a_{x,y}$ for $-\infty < x < +\infty$ with neighbours listed in the table:

| $y$ | Neighbours of $a_{x,y}$ with even $x, y$ |
|---|---|
| $-4$ | $a_{x-1,-3},\ a_{x,-2},\ a_{x+1,-3}$ (boundary) |
| $-2$ | $a_{x-1,-3},\ a_{x-1,-1},\ a_{x,0},\ a_{x+1,-1},\ a_{x+1,-3},\ a_{x,-4}$ |
| 0 | $a_{x-1,-1},\ a_{x-1,1},\ a_{x,2},\ a_{x+1,1},\ a_{x+1,-1},\ a_{x,-2}$ |
| $+2$ | $a_{x-1,1},\ a_{x-1,3},\ a_{x+1,3},\ a_{x+1,1},\ a_{x,0}$ (boundary) |

$\underset{\text{odd in } belt}{atom}(x, y)$ is a cell $a_{x,y}$ for $-\infty < x < +\infty$ with neighbours listed in the table:

| $y$ | Neighbours of $a_{x,y}$ with odd $x, y$ |
|---|---|
| $-3$ | $a_{x-1,-4},\ a_{x-1,-2},\ a_{x,-1},\ a_{x+1,-2},\ a_{x+1,-4}$ (boundary) |
| $-1$ | $a_{x-1,-2},\ a_{x-1,0},\ a_{x,1},\ a_{x+1,0},\ a_{x+1,-2},\ a_{x,-3}$ |
| $+1$ | $a_{x-1,0},\ a_{x-1,2},\ a_{x,3},\ a_{x+1,2},\ a_{x+1,0},\ a_{x,-1}$ |
| $+3$ | $a_{x-1,2},\ a_{x+1,1},\ a_{x,1}$ (boundary) |

Separation of cells with even and odd coordinates is justified by their exclusively hexagonal shape that enables tiling, and necessary because at points $(x, y)$ of different parity of coordinates, no cells exist. The inner and boundary atoms contain different number of nodes-cells.[6]

Now, let us take out certain cells from the belt's two horizontal boundaries and construct an auxiliary c-e structure of them called "*yjoint*", depicted in Fig. 3.2. When added to the *belt* in Fig. 3.1, the horizontal "joint" brings the blue and green $x$-directed boundaries of the *belt* to each other. This is a virtual "wrap-around" of the *belt* along axis $x$. Metaphorically–this summation transforms the *belt* into an infinite cylinder.
The c-e structure *yjoint* in Fig. 3.2, is the sum of its atomic cells:

$$yjoint = \underbrace{\sum_{-\infty<x<+\infty}\ \sum_{y\in\{-4,+2\}}}_{\text{for even } x,y} \underset{\text{even in } yjoint}{atom}(x, y) + \underbrace{\sum_{-\infty<x<+\infty}\ \sum_{y\in\{-3,+3\}}}_{\text{for odd } x,y} \underset{\text{odd in } yjoint}{atom}(x, y)$$

[6] In this example with the plane surface tiled wih hexagons, the inner atoms contain 7 cells (nucleus with neighbours), whereas those on the boundaries contain 4 or 6. So, such flat placed c-e structure is not cellular, but it will be complemented with an auxiliary structure, in order to get a cellular c-e structure.

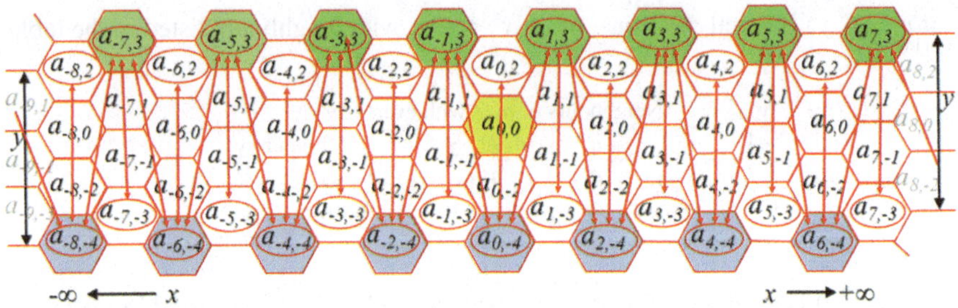


**Fig. 3.3** The flat view of outcome of joining the boundaries of the infinite belt shown in Fig. 3.1. Each cell has now 6 neighbours. Cellular c-e structure *infcylinder*

where $\underset{\text{even in } yjoint}{atom}(x, y)$ is a cell $a_{x,y}$ surrounded by neighbours listed in the table:

| $y$ | Neighbourhood of $a_{x,y}$ with even $x$, $y$ |
|---|---|
| $-4$ | $a_{x-1,3}$, $a_{x-1,-3}$, $a_{x+1,-3}$, $a_{x+1,3}$, $a_{x,2}$ |
| $+2$ | $a_{x-1,3}$, $a_{x,-4}$, $a_{x+1,3}$ |

and $\underset{\text{odd in } yjoint}{atom}(x, y)$ is a cell $a_{x,y}$ surrounded by neighbours listed in the table:

| $y$ | Neighbourhood of $a_{x,y}$ with odd $x$, $y$ |
|---|---|
| $-3$ | $a_{x-1,-4}$, $a_{x+1,-4}$, $a_{x,3}$ |
| $+3$ | $a_{x-1,2}$, $a_{x-1,-4}$, $a_{x,-3}$, $a_{x+1,-4}$, $a_{x+1,2}$ |

Summing the auxiliary, not cellular c-e structures *belt* and *yjoint*, the cellular c-e structure *infcylinder* is obtained:

$$infcylinder = belt + yjoint$$

Remember that the cells identically labelled in the summands, become one cell surrounded by the union of neighbourhoods of the summands. This follows from the idempotency of addition operation. The flattened image of *infcylinder* is in Fig. 3.3 and spatial in Fig. 3.4. The latter being an infinite cylindric base for a cellular c-e structure. The bidirectional red arrows linking cells from opposite boundaries in Fig. 3.3 show that after adding *yjoint* to *belt*, the cells pointed to by them–become neighbours. One may say that, this summation "implements" a transformation of the *belt* into infinite cylinder.[7]

[7] Note that the reverse transformation–the cylinder into the flat area–reminds the Mercator's geographic mapping. Obviously, all these transformations are not homeomorphic: they are not continuous. In our case–they do not preserve neighbourhoods. Our assumption on adjacency of sides (edges) of neighbouring cells is violated in the flat image of cylinder in Fig. 3.3. This assumption is restored as result of geometric transformation of this flat area into cylindric area shown in Fig. 3.4.

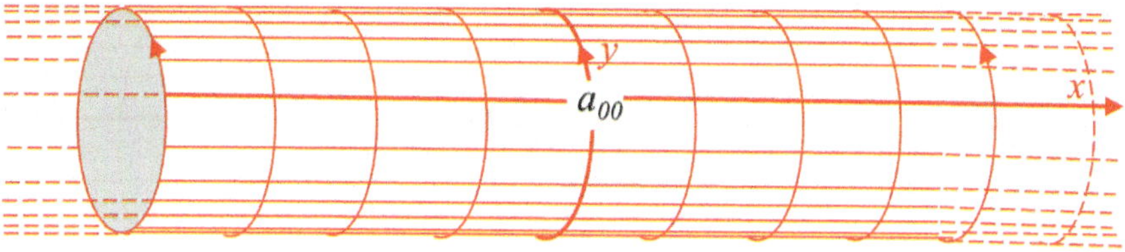


**Fig. 3.4** Visual result of bending bounded area of hexagonal cells shown in Fig. 3.1, around the $x$-axis, into the infinite cylinder. Algebraically defined as *infcylinder* = *belt* + *yjoint* which virtually realizes ("pretends") circular summation modulo 4–due to the idempotence of addition of c-e structures

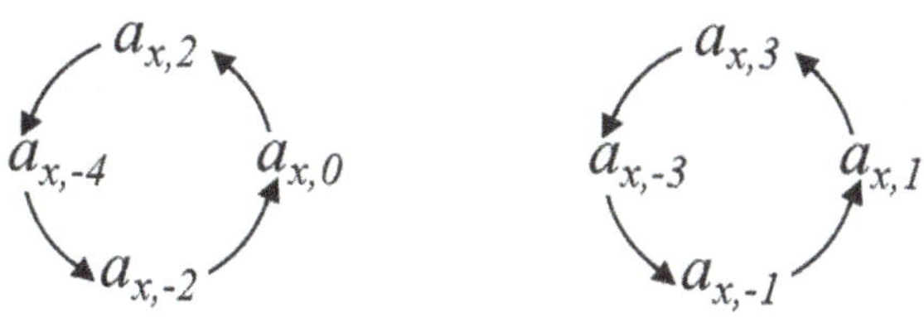


**Fig. 3.5** Cross-sections of infinite cylinder in Fig. 3.4 for even and odd $x$. The arrows indicate the successor/predecessor relationship of cells along axis $y$

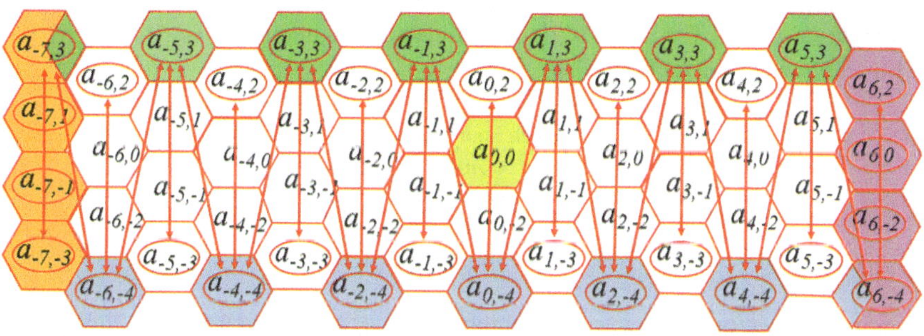


**Fig. 3.6** The c-e structure (non-cellular) named *cylinder*: an image of cut-off the area in Fig. 3.3 along vertical sides with different parity of coordinates

Geometric planar image of this transformation's outcome in Fig. 3.3 may be treated as 2-dimensional representation of infinite 3-dimensional cylinder depicted in Fig. 3.4. The cross-section of this cylinder is in Fig. 3.5. The columns from the planar image have been rolled up into circles.

To construct some curved surfaces–bases for finite cellular c-e structures like finite cylinder or torus–let us truncate the cellular c-e structure *infcylinder*, the area in Fig. 3.3, from both vertical sides with different parity of coordinates. This gives the bounded flat area *cylinder*, between vertical boundaries (orange and purple) and horizontal boundaries (blue and green) like, for instance, in Fig. 3.6. Its spatial view is in Fig. 3.7.

**Fig. 3.7** Spatial view of the c-e structure *cylinder* outspread on the flat area in Fig. 3.6

The bounded c-e structure *cylinder* in Fig. 3.6 differs from *infcylinder* only with the range of $x$ (and consequently–fewer neighbours of cells on vertical boundaries). This is the sum of its atomic cells:

$$cylinder = \underbrace{\sum_{-6\leq x\leq +6}\ \sum_{-4\leq y\leq +2}}_{\text{for even } x,y} \underset{\text{even in } cylinder}{atom}(x, y) + \underbrace{\sum_{-7\leq x\leq +5}\ \sum_{-3\leq y\leq +3}}_{\text{for odd } x,y} \underset{\text{odd in } cylinder}{atom}(x, y)$$

$\underset{\text{even in } cylinder}{atom}(x, y)$ for $-6 \leq x \leq +4$ neighbours of $a_{x,y}$ with even $x$, $y$ are the same as in *infcylinder*, while for $x = +6$ (right boundary), neighbours of $a_{6,y}$ (readily visible in Fig. 3.6) are in the table:

| $y$ | Neighbours of $a_{6,y}$ |
|---|---|
| $-4$ | $a_{5,-3}$, $a_{5,3}$, $a_{6,-2}$, $a_{6,2}$ |
| $-2$ | $a_{5,-3}$, $a_{5,-1}$, $a_{6,0}$, $a_{6,-4}$ |
| $0$ | $a_{5,-1}$, $a_{5,1}$, $a_{6,2}$, $a_{6,-2}$ |
| $+2$ | $a_{5,1}$, $a_{5,3}$, $a_{6,0}$, $a_{6,-4}$ |

$\underset{\text{odd in } cylinder}{atom}(x, y)$ for $-5 \leq x \leq +5$ neighbours of $a_{x,y}$ with even $x$, $y$ are the same as in *infcylinder*, while for $x = -7$ (left boundary), neighbours of $a_{-7,y}$ (readily visible in Fig. 3.6) are in the table:

| $y$ | Neighbours of $a_{-7,y}$ |
|---|---|
| $-3$ | $a_{-7,-1}$, $a_{-7,3}$, $a_{-6,-2}$, $a_{-6,-4}$ |
| $-1$ | $a_{-7,-1}$, $a_{-6,0}$, $a_{-6,-2}$, $a_{-7,-3}$ |
| $+1$ | $a_{-7,-3}$, $a_{-6,2}$, $a_{-6,0}$, $a_{-7,-1}$ |
| $+3$ | $a_{-7,-3}$, $a_{-6,2}$, $a_{-7,1}$, $a_{=6,-4}$ |

Now, let us take out certain cells from the vertical boundaries of *cylinder* and construct an auxiliary c-e structure of them called "*xjoint*", depicted in Fig. 3.8. When added to the *cylinder* in Fig. 3.6, the vertical "joint" brings the orange and pink $y$-directed boundaries of the *cylinder* to each other. This transforms the *cylinder* into a *torus*.

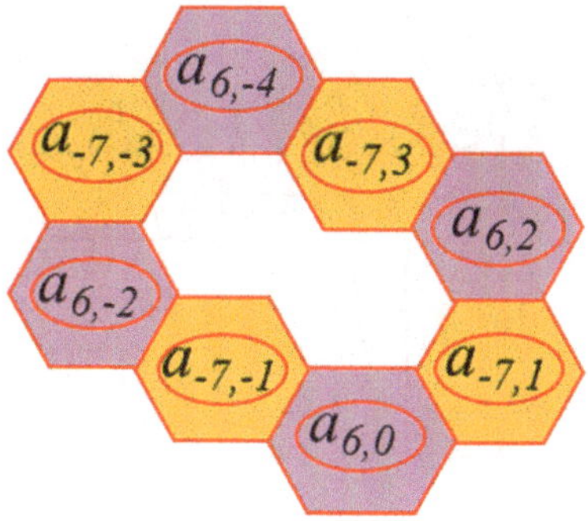


**Fig. 3.8** C-e structure *xjoint*. When added to *cylinder*–that in Fig. 3.6–the image of torus arises, flatten (outspread) in Fig. 3.9 and spatial in Fig. 3.10

The c-e structure *xjoint* in Fig. 3.8 is the sum of its atomic cells:

$$xjoint = \sum_{y\in\{-4,-2,0,+2\}} \underset{\text{even in } xjoint}{atom}(6, y) + \sum_{y\in\{-3,-1,+1,+3\}} \underset{\text{odd in } xjoint}{atom}(-7, y)$$

$\underset{\text{even in } xjoint}{atom}(6, y)$ is a cell $a_{x,y}$ surrounded by neighbours listed in the table:

| $y$ | Neighbourhood of $a_{6,y}$ |
|---|---|
| $-4$ | $a_{-7,-3}$, $a_{-7,3}$ |
| $-2$ | $a_{-7,-3}$, $a_{-7,-1}$ |
| 0 | $a_{-7,-1}$, $a_{-7,1}$ |
| $+2$ | $a_{-7,1}$, $a_{-7,3}$ |

$\underset{\text{odd in } xjoint}{atom}(-7, y)$ is a cell $a_{x,y}$ surrounded by neighbours listed in the table:

| $y$ | Neighbourhood of $a_{-7,y}$ |
|---|---|
| $-3$ | $a_{6,-4}$, $a_{6,-2}$ |
| $-1$ | $a_{6,-2}$, $a_{6,0}$ |
| $+1$ | $a_{6,0}$, $a_{6,2}$ |
| $+3$ | $a_{6,2}$, $a_{6,-4}$ |

The specification of the cellular c-e structure named *torus* outspread on the toroidal surface is the sum:

$$torus = cylinder + xjoint$$

Its flattened view is in Fig. 3.9 and its spatial view in Fig. 3.10.

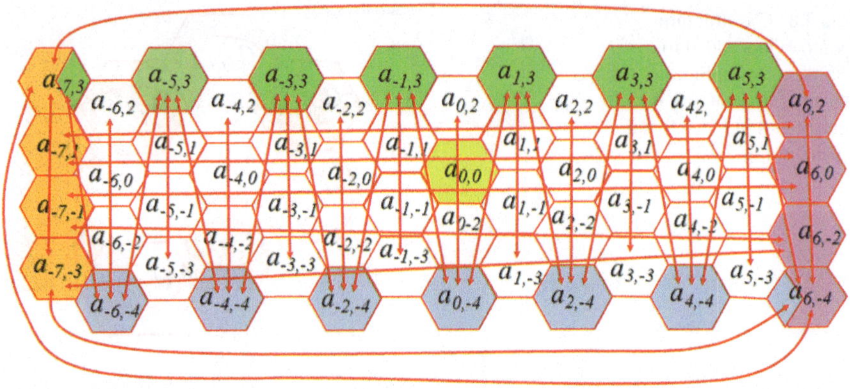


**Fig. 3.9** The pictorial outcome of joining the vertical boundaries of *cylinder*: the flatten view of *torus*. Cellular c-e structure–each cell has 6 neighbours

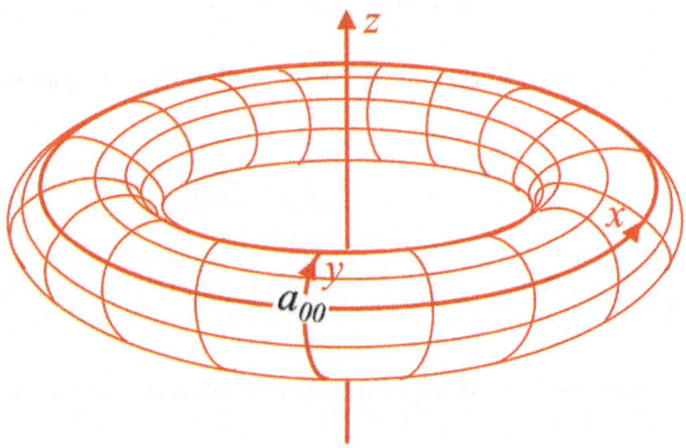


**Fig. 3.10** Torus, a visual result of circular bending the bounded area depicted in Fig. 3.9 along both directions $x$ and $y$ (or if someone prefers–bending the finite cylinder in Fig. 3.7 around the axis $z$)

## 3.3 Concluding Remarks

1. In the previous chapters 2-dimensional cellular c-e structures were considered. The main attempt to exploit cause-effect structures for some issues in cellular automata, was turning the geometric base of cellular c-e structures–the Cartesian planar surface–into some curved surfaces. To this end a set of atomic c-e structures was singled out of entire plane, arranging them along one or both axes into a "belt" or "array" (infinite or finite). Then, addition operation in the quasi semiring of c-e structures was applied to bring together some boundaries of the "belt". This simulated a distortion of the "belt", as if it were bent into a surface of an object in 3-dimensional space. This procedure required assumption on the polygonal shape of the cells evenly distributed on the surface. For examples of this procedure, cells with six neighbours have been taken, thus of hexangular shape. This seems more challenging than in case of square (rectangular more generally) shape of cells, because joining different sides of the "belt" needs equinumerous neighbourhoods

of cells in resulted cellular c-e structure, thus cells on bounderies coming together should have different parity of coordinates. What's more, not in all points $(x, y)$ the cells exist: only in those with the same parity of $x$ and $y$.

2. It was mentioned in a note in Sect. 3.1 that the 2-dimensional surface of a ball in 3-dimensional space cannot be covered only by hexagons. Thus no applicable procedure exists. In particular, the one applied above for cylinders and toruses. Simply, there is no way to join boundaries of a polygon (rectangle in case here used) covered with hexagons sharing sides, so that on the joined boundaries of the polygon, the hexagons would "fit" one another on the resulting sphere, preserving adjacency of sides. Formally, this fact directly follows from the well-known Euler formula [2, 3]: $\chi = V - E + F$ characterizing a class (genus) of surfaces of 3-dimensional polyhedra, where $V$ is a number of vertices, $E$–of edges, $F$–of faces, see references on algebraic topology. For the surface of convex polyhedra in 3-dimensional space as well as for their homeomorphic counterparts (in particular surfaces of 3-dimensional balls), the Euler characteristic equals 2. A simple spatial imagination brings to conclusion that no convex polyhedron with all hexagonal faces exists,[8] thus no sphere circumscribed on such object–homeomorphic (topologically identical) to its surface. But counting of $V, E, F$ ensures this formally. Indeed, in each such polyhedron, every edge would belong to exactly two faces and every vertex to three, thus the polyhedron with $F$ hexagonal faces has $3F$ edges and $2F$ vertices, hence $\chi = 2F - 3F + F = 0$, so $\chi \neq 2$. Examples of Euler characteristic of some surfaces are: for cylinder, torus, Klein bottle, Möbius strip, $\chi = 0$, for double torus $\chi = -2$. Obviously, the tetrahedron, hexahedron (cube in particular), octahedron, dodecahedron, icosahedron are convex polyhedra, thus for them $\chi = 2$. The examples give some insight on feasibility of constructing curved bases for cellular c-e structures with required size of cells' neighbourhood. It was seen, for instance, that a cellular c-e structure with six neighbours of cells sharing sides cannot be placed on a sphere.
3. In above exemplary considerations a cut-off of rectangular "belt" (array) of cells from infinite Cartesian plane covered by a cellular c-e structure, has been realized along coordinates $x$ and $y$. Obviously, one may cut-off not necessarily a "belt", but another set of cells as a carrier of a c-e structure, which is to be laid on a surface curved–in effect of bending the plane. For instance a certain polygon. A planar form of this c-e structure is not a cellular c-e structure because cells on the boundary have less neighbours than those inside the cut-off. The equinumerous neighbourhood of all cells is restored as result of transformation of this planar cut-off fragment into a 2-dimensional closed surface (a manifold) of a 3-dimensional object. In our examples into the infinite cylinder and torus. In broad terms, the described way of transforming c-e structures from flat surfaces into cellular c-e structures on 2-dimensional manifolds (surfaces of solid figures), was prompted by the fact that every closed surface[9] is homeomorphic to

[8] Try to make a polyhedron with each face being a hexagon!

[9] A closed boundary of a solid figure, like sphere, torus, Klein bottle, etc. so not in the sense of the topological "closed set."

a topological space obtained by (pairwise) identifying sides of a certain polygon, of even number of sides. Since the identification may proceed point-by-point, in particular in different orientation of the sides, different closed surfaces may be obtained from the same polygon, e.g. a sphere or a Klein bottle from a rectangle. The attempt to exploit operations on cellular c-e structures to their spreading over the bent surfaces seems to give a certain (algorithmic) method to attain a desired spread. The kind of a bent surface depends on selection of cells from the polygon's boundary so that to make an auxiliary c-e structure of them (like those in Figs. 3.2, 3.8), which would complement (by summation) the c-e structure lying on the polygon. The outcome of the complementation is a desired cellular c-e structure on a bent surface.

4. In Sect. 3.2 on bending planar bases for cellular c-e structures, the hexagonal polygons–tiles representing cells have been used to illustrate the bending procedure which exploits the c-e structures summation. It was assumed that the hexagons share sides, so the plane (with Cartesian coordinates) were tiled with them. Obviously, there are only three types of such tessellations by means of identical polygons: with triangles, squares and hexagons. Figure 3.11 depicts a (piece of) plane surface hosting a c-e structure where each cell has three neighbours. Geometrically the plane is tessellated by equilateral triangles, so this is a triangulation of the plane. Since every closed 2-dimensional surface is triangulable,[10] such surface (and surfaces of convex polyhedra and their homeomorphic counterparts, like cylinders, spheres, toruses, etc.), may be a base for c-e structures composed of triangular cells sharing sides. But not necessarily preserving all features of cellular automata, as shows Fig. 3.11, where the neighbourhoods of cells with different parity of coordinates are not equivalent relative-to shift transformation, although are equinumerous.

   Atomic c-e structures shown in Fig. 3.12 with nuclei of the same coordinates parity are equivalent relative-to the shift transformation. The same concerns atomic c-e structures with nuclei of the different coordinates parity.

   Similarly to cellular c-e structures with rectangular and hexagonal cells-tiles (Figs. 3.1, 3.9 in Chap. 2), the c-e structure with triangular cells shown in Fig. 3.11 may be described algebraically as the sum of atoms shown in Fig. 3.12 as follows:

$$\sum_{-\infty<x<+\infty} \sum_{-\infty<y<+\infty} \underset{\text{equal parity of } x,y}{atom3(x,y)} + \sum_{-\infty<x<+\infty} \sum_{-\infty<y<+\infty} \underset{\text{different parity of } x,y}{atom3(x,y)}$$

[10] In dimensions not greater than 3, every manifold is triangulable. The triangle–2-dimensional simplex, a basic building block for more complicated structures like two-dimensional complexes, the polyhedra in particular, underlies a certain feature of topological spaces: those homeomorphic to surfaces of polyhedra (i.e. built of triangles). Such space is called triangulable.

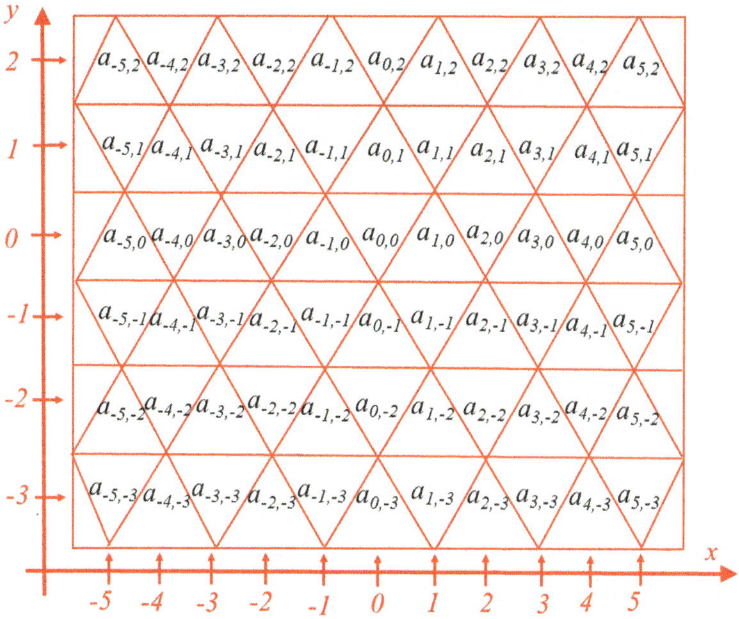


**Fig. 3.11** Covering the Cartesian plane with triangular cells sharing sides with neighbours. Atomic c-e structures with nuclei of the same coordinates parity only, are equivalent (among themselves) relative-to the shift transformation, as well as only those of the different parity. But not all among themselves, so such c-e structure does not fulfil requirement of the cellular automata

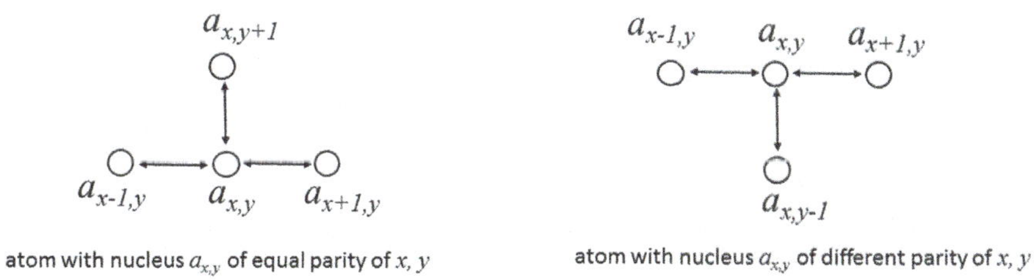


**Fig. 3.12** Atomic c-e structure with nucleus $a_{xy}$ and neighbourhood of 3 cells: $\underbrace{atom3(x, y)}_{\text{equal parity of } x,y} = [a_{x,y}, \{a_{x-1,y}, a_{x,y+1}, a_{x+1,y}\}]$ $\underbrace{atom3(x, y)}_{\text{different parity of } x,y} = [a_{x,y}, \{a_{x-1,y}, a_{x+1,y}, a_{x,y-1}\}]$

# References

1. Dold, A.: Lectures on Algebraic Topology
2. Euler Characteristic: https://en.wikipedia.org/wiki/Euler_characteristic#Plane_graphs
3. Euler Leonhard, "Elementa doctrinae solidorum". Novi Commentarii Academiae Scientiarum Petropolitanae: 109–140, 1758
4. Hatcher, A.: Algebraic Topology. Cambridge University Press (2002)

## References

# Chapter 4
# Beyond the 2-Dimensional Bases

There is an essential difference between cellular c-e structures outspread over multi-dimensional and one-dimensional bases (location spaces). In regard to structural, that is geometric features, any cell in a cellular c-e structure on a 2-dimensional base or of higher dimension, may have arbitrary (but equal for all cells) number of neighbours. The c-e structures on 1-dimensional bases, e.g. on infinite straight line, on a circle or on their homeomorphic counterparts, are composed of cells having only two neighbours. The procedure for cutting a segment from a straight line containing the cellular c-e structure and then "bending" this segment into a closed curve by making its ends adjoined, so that they become neighbours, is simpler than in the 2-dimensional cases considered in the previous chapter. The analogy to 2-dimensional geometrical arrangement of polygonal cells whose adjacency means that they are neighbours, in 1-dimensional case amounts to contiguity of their coordinates. Figure 4.1 shows a segment of a c-e structure cut out from the infinite cellular c-e structure on the straight line, the auxiliary c-e structure with cells taken from the left and right end of the segment, and the result of summing up of these two (not cellular) structures into the cellular one arranged in a circle. An atom of this infinite 1-dimensional cellular c-e structure is $\overset{a_x}{a_{x-1}} \leftrightarrow \overset{a_{x-1}+a_{x+1}}{a_x} \leftrightarrow \overset{a_x}{a_{x+1}}$, with the cause and (the same) effect polynomials above names of cells.[1] So the algebraic notation ("arrow expression"–see Chap. 1) of the infinite structure is:

$$\sum_{-\infty<x<\infty} \overset{a_x}{a_{x-1}} \leftrightarrow \overset{a_{x-1}+a_{x+1}}{a_x} \leftrightarrow \overset{a_x}{a_{x+1}}.$$

[1] Attention: distinguish "+" as addition of nodes of c-e structures (as in $a_x + a_y$) from addition of numbers (as in $a_{x+y}$).



L. Czaja, *Cellular Cause-Effect Structures*, Studies in Systems, Decision and Control 526,
https://doi.org/10.1007/978-3-031-55590-9_4

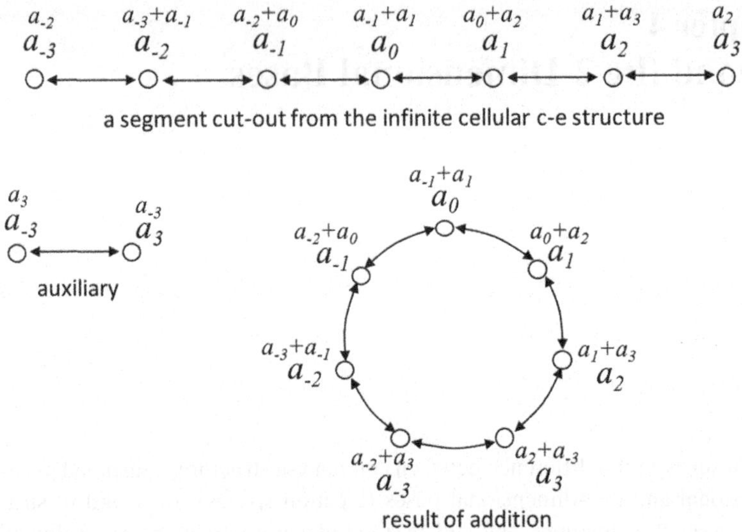


**Fig. 4.1** Result of summation of the auxiliary c-e structure and the segment (not cellular). Circular-shape of this result is a cellular c-e structure. The cause/effect polynomials are written over the names of cells

Its segment in Fig. 4.1 is:

$$\sum_{-2\leq x\leq +2} \overset{a_x}{a_{x-1}} \leftrightarrow \overset{a_{x-1}+a_{x+1}}{a_x} \leftrightarrow \overset{a_x}{a_{x+1}}.$$

Thus, the sum

$$\overset{a_3}{a_{-3}} \leftrightarrow \overset{a_{-3}}{a_3} + \sum_{-2\leq x\leq +2} \overset{a_x}{a_{x-1}} \leftrightarrow \overset{a_{x-1}+a_{x+1}}{a_x} \leftrightarrow \overset{a_x}{a_{x+1}}$$

is the cellular c-e structure of circular shape in Fig. 4.1.

In regard to behavioural features, the formula of semantics for 1-dimensional cellular c-e structure $U$ is analogous to (B1D1) in Chap. 2 (section 3) but with one coordinate $x$ of the cell names:

$$(s,t) \in [\![ U ]\!] \overset{\text{def}}{\Leftrightarrow} \forall a_x \in car(U) : (s,t) \in [\![ a_x ]\!]$$

where now $(s,t) \in [\![ a_x ]\!]$ holds if and only if:

$$a_x \in t \wedge [(a_x \notin s \wedge nbh(a_x) \cap s \in B1) \vee (a_x \in s \wedge nbh(a_x) \cap s \in D1)]$$

with $nbh(a_x) = \{a_{x-1}, a_{x+1}\}$ for $-\infty < x < +\infty$ and $nbh(a_x) = \{a_{x\ominus 1}, a_{x\oplus 1}\}$ for $p \leq x \leq q$ where $p, q$ are extreme coordinates of a segment cut off from the infinite cellular c-e structure ($p = -3$, $q = +3$ in Fig. 4.1), and $\ominus$, $\oplus$ are subtraction and addition of integer numbers modulo $q - p + 1$ thus:

$$x \oplus 1 = \begin{cases} x + 1 & \text{if } p \le x \le q - 1 \\ p & \text{if } x = q \end{cases}$$

$$x \ominus 1 = \begin{cases} x - 1 & \text{if } p + 1 \le x \le q \\ q & \text{if } x = p \end{cases}$$

As everywhere so far, any cell assumes status ***true*** or ***false***, $s$, $t \subseteq car(U)$ are a current and next-state of $U$ and $car(U)$ is the set of all nodes-cells of $U$.[2] Remind: $nbh(a_x) \cap s$ is the set of active (assuming ***true***) neighbours of $a_x$ in the state $s$. For a cell $a_x$, the parameters $B1$ and $D1$ are now families of subsets of $nbh(a_x)$, that is $B1 \subseteq 2^{nbh(a_x)}$ and $D1 \subseteq 2^{nbh(a_x)}$ and, as in the 2-dimensional case, they specify behaviour of $U$, so they concretize the schematic semantics of this cellular c-e structure. Since now $|nbh(a_x)| = 2$, the number of local next-state rules of 1-dimensional cellular c-e structures is $2^{2^3} = 256$. Every rule is determined by parameters $B1$, $D1$. For instance, if $B1 = \{\{a_{x-1}, a_{x+1}\}, \{a_{x+1}\}\}$ then $a_x$ changes its passive ($a_x \notin s$) status into active ($a_x \in t$) iff $nbh(a_x) \cap s \in B1$, whereas if $D1 = \{\{a_{x-1}\}, \{a_{x+1}\}, \{\}\}$ then $a_x$ active ($a_x \in s$) in $s$ remains active in $t$ iff $nbh(a_x) \cap s \in D1$.[3] Obviously, this concerns infinite cellular c-e structures. In their finite circular shape like in Fig. 4.1, $a_{x-1}$, $a_{x+1}$ should be replaced with $a_{x\ominus 1}$, $a_{x\oplus 1}$.

An example of evolution of the circular shaped cellular c-e structure in Fig. 4.1 starting from the initial state $s_0 = \{a_{-2}, a_0\}$ depicts Fig. 4.2.

**Remark**

The parameters $B1$, $D1$ for 1-dimensional cellular c-e structures with 2-valued cells, express the same as the so-called "Rule $R(a_{x-1}, a_x, a_{x+1})$", a decimally coded next-state rule appearing in writings on cellular automata. Every such rule $R$ is an arithmetical representation of a three argument Boolean function, thus the parameters $B1$, $D1$ represent such a Boolean function in terms of the cellular c-e structures. Net-like representation of Boolean functions in the form of the cause-effect structure graphs or expressions, may be found in [2]. Since there are a number of representations of such functions (like truth tables, propositional formulas, algorithmic [3], Boolean networks [6], in programming languages [1]), one may immediately write parameters $B1$, $D1$ for a given representation of a Boolean function. As an example, let us write the parameters $B1$, $D1$ for 3-argument threshold function [4] represented by the formula $(a_{x-1} \wedge a_x) \vee (a_{x-1} \wedge a_{x+1}) \vee (a_x \wedge a_{x+1})$.[4] If in a state $s$ the cell

[2] The 1-dimensional cellular automata with 2 values of state are usually called "elementary". Since this name is used in different sense for cause-effect structures, here it is avoided. By the way: such seemingly simple, one-dimensional cellular c-e structures have expressive abilities like Turing machines.

[3] The parameters $B1 = \{\{a_{x-1}, a_{x+1}\}, \{a_{x+1}\}\}$, $D1 = \{\{a_{x-1}\}, \{a_{x+1}\}, \{\}\}$ define the so-called "rule 110" in writings on cellular automata.

[4] The orthographic non-uniformity of using "$a_{x-1}$, $a_x$, $a_{x+1}$" as propositional variables in this formula and also as names of cells belonging or not to a state $s$, should not mislead: $a_x \in s$ means $a_x$ has value ***true*** in $s$, etc.

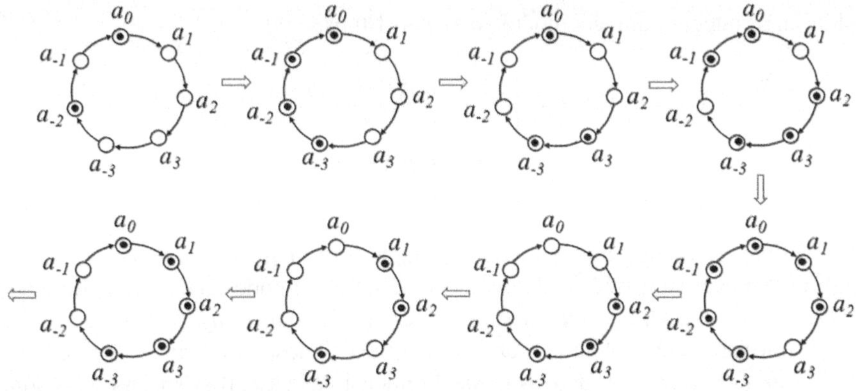


**Fig. 4.2** Eight successive stages generating states $s_0, s_1, s_2, \ldots, s_7$ of evolution of 1-dimensional cyclic cellular c-e structure from Fig. 4.1, with parameters $B1 = \{\{a_{x\ominus 1}, a_{x\oplus 1}\}, \{a_{x\oplus 1}\}\}$, $D1 = \{\{a_{x\ominus 1}\}, \{a_{x\oplus 1}\}, \{\}\}$, starting from the initial state $s_0 = \{a_{-2}, a_0\}$. Note that if the initial state were $\{a_{-3}, a_{-2}, a_{-1}, a_0, a_1, a_2, a_3\}$ (tokens at all nodes), then the next-state (empty nodes) would be terminal

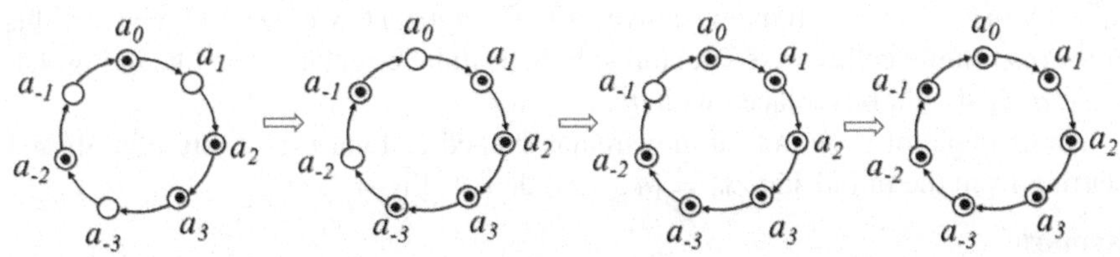


**Fig. 4.3** Four successive stages generating states $s_0, s_1, s_2, s_3$ of evolution of 1-dimensional cellular c-e structure "cyclic threshold", with parameters $B1 = \{\{a_{x\ominus 1}, a_{x\oplus 1}\}\}$, $D1 = \{\{a_{x\ominus 1}\}, \{a_{x\oplus 1}\}, \{a_{x\ominus 1}, a_{x\oplus 1}\}\}$, starting from the initial state $s_0 = \{a_{-2}, a_0, a_2, a_3\}$. The evolution stabilizes at state $s_3$–every next-state is $s_3$

$a_x$ holds ***false*** ($a_x \notin s$) then $a_x$ will hold ***true*** ($a_x \in t$) in the next-state $t$, what takes place iff $B1 = \{\{a_{x-1}, a_{x+1}\}\}$. The cell $a_x$ with ***true*** in the state $s$ does not change status in the next-state $t$ iff $D1 = \{\{a_{x-1}\}, \{a_{x+1}\}, \{a_{x-1}, a_{x+1}\}\}$. Evolution of the circular shaped cellular [modulo $3-(-3)+1$] c-e structure "threshold" from Fig. 4.1, starting from initial state $s_0 = \{a_{-2}, a_0, a_2, a_3\}$, depicts Fig. 4.3.

Now, let us consider 3-dimensional cellular c-e structures, thus composed of cells labelled by $a_{x,y,z}$. Geometrically, the cells will be polyhedrons (in particular prisms), and as in the 2-dimensional case, the neighbours of a cell surround it as closely as possible. Some of the challenge is to arrange the cells so that they share faces, and the 3-dimensional space is filled with identical cells–a kind of tessellation of space with identical polyhedrons. This is not always possible, but it can be done e.g. with cubic cells that share common faces, thus covering the space, as in Fig. 4.4.

As in the 2-dimensional case, where adjacency of sides between cells and their neighbours may be geometrically expedient for some applications though immaterial for description of semantics, all the more it concerns 3-dimensional cases, where

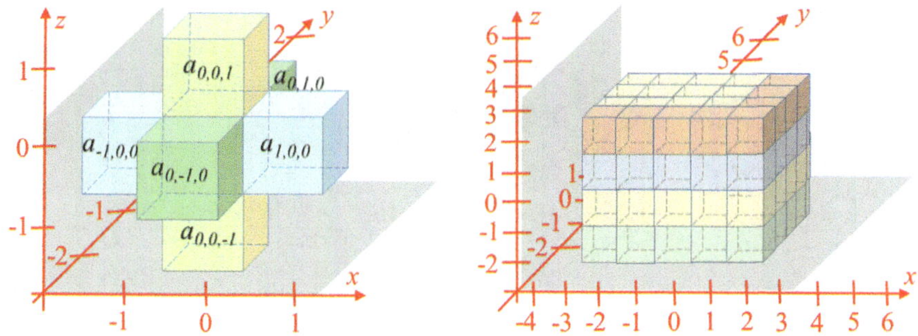


**Fig. 4.4** Atomic cellular c-e structure with nucleus $a_{0,0,0}$ hidden between its cubic neighbours. Also a piece of space covered by cubic cells with adjacent faces. Thus, for every cell $a_{x,y,z}$, $nbh(a_{x,y,z}) = \{a_{x-1,y,z}, a_{x,y+1,z}, a_{x+1,y,z}, a_{x,y-1,z}, a_{x,y,z-1}, a_{x,y,z+1}\}$

polyhedrons (prisms as a special case) replace polygons and faces replace sides.[5] Thus, if we want any cell to connect to its neighbours only through its faces, there should be a one-to-one correspondence between the faces of the cell and its neighbours. This requires identical type and size of polyhedrons. Otherwise, the neighbourhoods of all cells would not be equinumerous. Moreover, if infinite 3-dimensional space is to be covered ("tessellated") by disjoint identical and convex polyhedrons that share common faces, the problem of being able to do so arises. The polyhedrons are classified into some kinds, e.g. the so-called "Platonic", "Archimedean" (obtained by various ways of truncating corners in the Platonic), prism-shaped, pyramid-shaped regular, convex or not, and many others. The choice of a kind depends on a natural phenomena to be simulated by a cellular c-e structure. Of the five existent Platonic polyhedrons: tetrahedron, cube, octahedron, dodecahedron, icosahedron of 4, 6, 8, 12, 20 faces respectively, only cubes can fill the space without gaps, as shown in Fig. 4.4.[6] But the problem of covering space by polyhedral cells occurs also when they are cubes.

Let us consider a problem of 3-dimensional cellular c-e structures with polyhedral cells and identically behaving 2-dimensional with polygonal cells, and their capability to cover space and plane respectively. Recall that the cells are not in the vertices (corners) of these geometric figures but are these figures themselves,[7] and that neighbours of each cell are similarly ordered (a "shift equivalence of the atoms"). The cellular c-e structure $U_3$ in Fig. 4.5 with cubic cells in 3-dimensional space could behave like the cellular c-e structure $U_2$ with hexagonal cells, if a one-to-one mapping between the hexagons in $U_2$ and the cubes in $U_3$ is established, such that hexagons

[5] For instance, the geometrical shape of cells may be expedient for application in crystallography, in protein modelling and in a number of other fields.

[6] Notice that each cube shares 6 faces with neighbouring cubes in tessellation of space, just as each hexagon shares 6 sides with neighbouring hexagons in tessellation of plane surface shown earlier.

[7] Notice that this geometric perception does not contradict the purely graph-like one, where a cell is perceived as a graph vertex, thus a point inside a polygon or polyhedron.

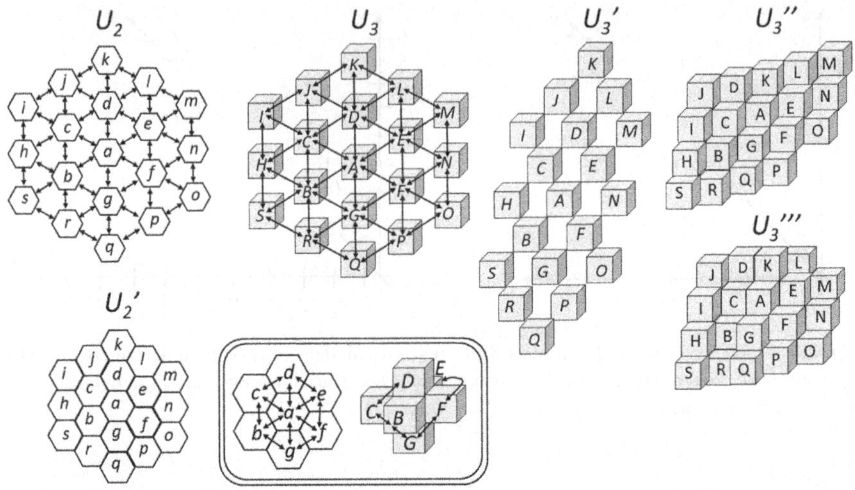


**Fig. 4.5** Fragments of 2-dimensional and 3-dimensional isomorphic cellular c-e structures, variously located in space. The bidirectional arrows link neighbouring cells in $U_2$ as well as neighbouring cells in $U_3$, related to the latter. In the boxed image, the hexagonal cells share sides with their neighbours, while the corresponding cubic cells cannot share faces with their neighbours if isomorphism between these atoms is to take place

$\varphi$, $\varphi'$ are neighbours if and only if their cubic images $\Phi$, $\Phi'$ are neighbours. So the neighbours of each hexagonal cell in $U_2$ and neighbours of each related cubic cell in $U_3$ are ordered in the same succession.[8] This means the isomorphism of graphs (cf [5]) between $U_2$ and $U_3$. Such isomorphism would ensure identical behaviour of $U_3$ and $U_2$, assuming the same structure of parameters $B1$, $D1$ appearing in the formula ($B1D1$) in Sect. 2.3 (Chap. 2) for these 3-dimensional and 2-dimensional arrangements of cells (with a suitable redefinition of state for the 3-dimensional case, that is assuming the same propositional values for the cells in $U_3$ and related to them in $U_2$).[9]

In $U_2'$, functionally identical to $U_2$, each cell shares sides with neighbours. In $U_3$, $U_3'$, $U_3''$ the cells do not share faces with neighbours. In $U_3'''$ some do, some do not. The hexagons marked by lower case letters and cubes by respective upper case, are in the one-to-one relationship. However none of the spatial arrangements preserving isomorphism with the planar $U_2'$ can cover the space without gaps if the cubic cells are to share faces with their neighbours. Such tessellation of space and retainment of this isomorphism (thus behaviour) are in contradiction.

---

[8] Establishing such a mapping takes into account the position of these polygons in the plane and polyhedrons in space, determined by their Cartesian coordinates. In the geometric perception promoted here, each side of a hexagonal cell in $U_2$ represents one neighbour of that cell, just as each face of a cubic cell in $U_3$ represents its one neighbour.

[9] This means $\varphi \in s_{U_2} \Leftrightarrow \Phi \in s_{U_3}$ for all related cells $\varphi$ and $\Phi$ in $U_2$ and $U_3$, where $s_{U_2}$, $s_{U_3}$ denote states of $U_2$ and $U_3$ respectively. Remember that $\varphi \in s_{U_2}$ means that the cell $\varphi$ assumes value ***true*** in the state $s_{U_2}$, etc.

In symbolic terms let a cubic cell $\Phi$ in $U_3$ marked by upper case $A, B, C, \ldots$ be in a one-to-one relationship with hexagonal cell $\varphi$ in $U_2$ marked by corresponding lower case $a, b, c, \ldots$ and let $s_{U_3}, t_{U_3}$ and $s_{U_2}, t_{U_2}$ denote states of $U_3$ and $U_2$ respectively. Let the following next-state rules $(B1_{U_3}D1_{U_3})$ and $(B1_{U_2}D1_{U_2})$ for $U_3$ and $U_2$ be admitted:

$$\left.\begin{array}{r}\Phi \in t_{U_3} \wedge \left[\Phi \notin s_{U_3} \wedge nbh(\Phi) \cap s_{U_3} \in B1_{U_3} \vee \right.\\ \left.\Phi \in s_{U_3} \wedge nbh(\Phi) \cap s_{U_3} \in D1_{U_3}\right]\end{array}\right\} \qquad (B1_{U_3}D1_{U_3})$$

$$\left.\begin{array}{r}\varphi \in t_{U_2} \wedge \left[\varphi \notin s_{U_2} \wedge nbh(\varphi) \cap s_{U_2} \in B1_{U_2} \vee \right.\\ \left.\varphi \in s_{U_2} \wedge nbh(\varphi) \cap s_{U_2} \in D1_{U_2}\right]\end{array}\right\} \qquad (B1_{U_2}D1_{U_2})$$

where $B1_{U_3} \subseteq 2^{nbh(\Phi)}$, $D1_{U_3} \subseteq 2^{nbh(\Phi)}$ are families of subsets of $nbh(\Phi)$ and similarly $B1_{U_2} \subseteq 2^{nbh(\varphi)}$, $D1_{U_2} \subseteq 2^{nbh(\varphi)}$ are families of subsets of $nbh(\varphi)$ (see the formula $(B1D1)$ in Sect. 2.3 (Chap. 2)). Remember that $nbh(\Phi) \cap s_{U_3}$ and $nbh(\varphi) \cap s_{U_2}$ are sets of active neighbours of the cells $\Phi$ and $\varphi$ respectively, in the states $s_{U_3}$, $s_{U_2}$. Thus, if $nbh(\Phi) \cap s_{U_3} \in B1_{U_3} \Leftrightarrow nbh(\varphi) \cap s_{U_2} \in B1_{U_2}$ and $nbh(\Phi) \cap s_{U_3} \in D1_{U_3} \Leftrightarrow nbh(\varphi) \cap s_{U_2} \in D1_{U_2}$ then $(s_{V_3}, t_{V_3}) \in [\![\, \Phi \,]\!]$ holds iff $(s_{V_2}, t_{V_2}) \in [\![\, \varphi \,]\!]$ provided that $\Phi \in s_{V_3} \Leftrightarrow \varphi \in s_{V_2}$ and $\Phi \in t_{V_3} \Leftrightarrow \varphi \in t_{V_2}$ for all related cells $\Phi$ and $\varphi$. Therefore $(s_{V_3}, t_{V_3}) \in [\![\, U_3 \,]\!]$ holds if and only if $(s_{V_2}, t_{V_2}) \in [\![\, U_2 \,]\!]$.[10]

For another example, let us take the hexagonal regular prism as the cell of a cellular c-e structure. Its atom and a fragment of space filled with its cells is in Fig. 4.6.

This example is intended to illustrate the identical behaviour of the 3-dimensional cellular c-e structure and its 2-dimensional counterpart, in this case for the Moore type Game of life cellular c-e structures. In the 2-dimensional version of this task (Fig. 2.6 in Chap. 2), the octagonal cells cannot cover the plane. The 3-dimensional space can be covered with hexagonal prisms,[11] if their special position is not required. But such tessellation is not possible if the prisms which stand for neighbouring cells in the c-e structure, should be arranged so that to share faces. The cellular c-e structure $V_3$ in Fig. 4.7 with prismatic cells in 3-dimensional space could behave like the cellular c-e structure $V_2$ with octogonal cells, if the isomorphism similar to that between $U_2$ and $U_3$ in Fig. 4.5 were established between the graphs $V_2$ and $V_3$, but different next-state rule were applied.

To express this more precisely, let a prismatic cell $\Gamma$ in $V_3$ marked by upper case $A, B, C, \ldots$ be in a one-to-one relationship with octagonal cell $\gamma$ in $V_2$ marked by corresponding lower case $a, b, c, \ldots$ and let $s_{V_3}, t_{V_3}$ and $s_{V_2}, t_{V_2}$ denote states of $V_3$

10 Notice that the parameters $B1_{U_3}$ and $B1_{U_2}$ are of the same structure, that is they differ only by related (i.e. isomorphic) names of cells they contain. These names (labels) represent the coordinates at which the cells are located. For instance, in the task the "two light rays" (Sect. 2.3 in Chap. 2), $B1_{U_2'} = \{\{\varphi_{x+1,y-1}\}, \{\varphi_{x-1,y+1}\}\}$, $D1_{U_2'} = \{\{\}, \{\varphi_{x+1,y-1}\}, \{\varphi_{x-1,y+1}\}\}$, and $B1_{U_3'} = \{\{\Phi_{x+1,y-1,z-1}\}, \{\Phi_{x-1,y+1,z+1}\}\}$, $D1_{U_3'} = \{\{\}, \{\Phi_{x+1,y-1,z-1}\}, \{\Phi_{x-1,y+1,z-1}\}\}$, where $\varphi_{x,y}$ is the name of the cell $\varphi$ located at point $(x, y)$ of the plane (in accordance with the arrangement of hexagonal cells presented by $U_2'$ in Fig. 4.5, the same as in Fig. 2.9 in Chap. 2). Similarly, $\Phi_{x,y,z}$ is the name of the cell $\Phi$ located at point $(x, y, z)$ of the space (in accordance with the arrangement of cubic cells presented by $U_3'$ in Fig. 4.5.

11 A kind of octahedral solids, not to be confused with Platonic octahedrons.

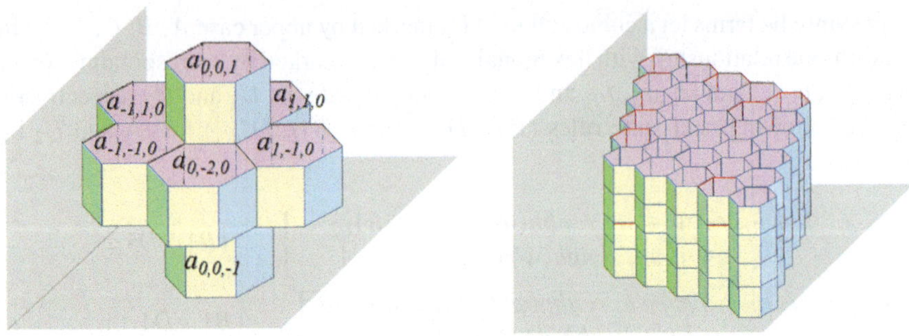


**Fig. 4.6** Atomic cellular c-e structure with nucleus $a_{0,0,0}$ hidden between its eight hexagonal prisms neighbours. The invisible cell $a_{0,2,0}$ is shielded by $a_{0,0,1}$. Also a piece of space covered by prismatic cells with adjacent faces. Thus, for every cell $a_{x,y,z}, nbh(a_{x,y,z}) = \{a_{x-1,y-1,z}, a_{x-1,y+1,z}, a_{x,y+2,z}, a_{x+1,y+1,z}, a_{x+1,y-1,z}, a_{x,y-2,z}, a_{x,y,z-1}, a_{x,y,z+1}\}$

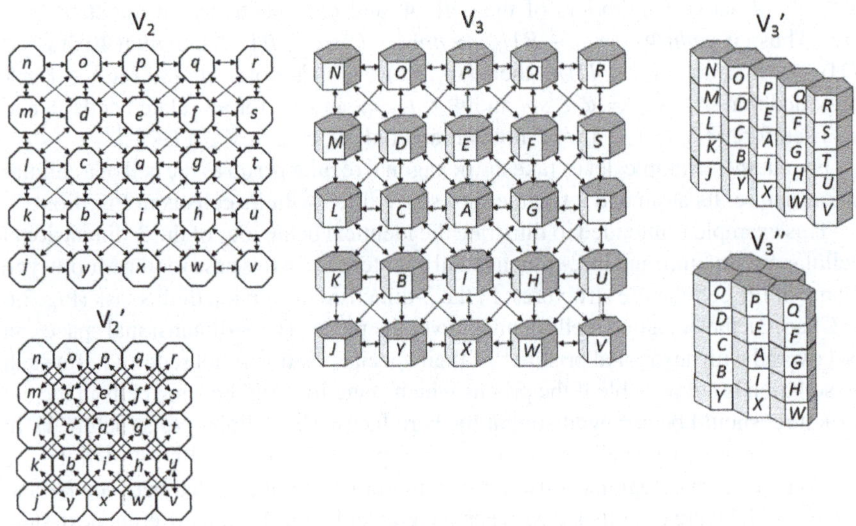


**Fig. 4.7** Fragments of cellular c-e structures: 2-dimensional with octagonal cells, and 3-dimensional with regular hexagonal prismatic cells, variously located in space. The comments from the previous example are also relevant to this example, with provision that here neither planar nor spatial version can tessellate the plane and space along with preserving isomorphism between these two versions, thus if they are to behave identically

and $V_2$ respectively. Let the following next-state rules $(B_{V_3} D_{V_3})$ and $(B_{V_2} D_{V_2})$ for $V_3$ and $V_2$ be be admitted:

$$\left.\begin{array}{r}\Gamma \in t_{V_3} \wedge \left[\Gamma \notin s_{V_3} \wedge |nbh(\Gamma) \cap s_{V_3}| \in B_{V_3} \vee \right. \\ \left. \Gamma \in s_{V_3} \wedge |nbh(\Gamma) \cap s_{V_3}| \in D_{V_3}\right]\end{array}\right\} \quad (B_{V_3} D_{V_3})$$

$$\left.\begin{array}{r}\gamma \in t_{V_2} \wedge \left[\gamma \notin s_{V_2} \wedge |nbh(\gamma) \cap s_{V_2}| \in B_{V_2} \vee \right. \\ \left. \gamma \in s_{V_2} \wedge |nbh(\gamma) \cap s_{V_2}| \in D_{V_2}\right]\end{array}\right\} \quad (B_{V_2} D_{V_2})$$

Thus, if $B_{V_3} = B_{V_2}$ and $D_{V_3} = D_{V_2}$ then $(s_{V_3}, t_{V_3}) \in [\![ \Gamma ]\!]$ holds if and only if $(s_{V_2}, t_{V_2}) \in [\![ \gamma ]\!]$ provided that $\Gamma \in s_{V_3} \Leftrightarrow \gamma \in s_{V_2}$ and $\Gamma \in t_{V_3} \Leftrightarrow \gamma \in t_{V_2}$ for all related cells $\Gamma$ and $\gamma$. Therefore $(s_{V_3}, t_{V_3}) \in [\![ V_3 ]\!]$ holds if and only if $(s_{V_2}, t_{V_2}) \in [\![ V_2 ]\!]$. If $B_{V_3} = B_{V_2} = \{3\}$, $D_{V_3} = D_{V_2} = \{2, 3\}$ then $V_3$ and $V_2$ perform the Moore type Game of life. So, their dimension and collocation of cells on the plane and space is immaterial for behaviour: $\gamma$ and $\Gamma$ are just different names of associated cells. But sharing faces of hexagonal prisms with their neighbours cannot be attained if the isomorphism is required between $V_3$ and $V_2$, thus identical behaviour. The tessellation of space by the prisms, though possible geometrically, precludes the isomorphism.

**Summarizing Remarks**

1. Identical behaviour of cellular c-e structures with hexagonal and cubic cells considered in the example shown in Fig. 4.5 assumed the existence of a one-to-one correspondence between cells in $U_2$ and $U_3$ and a similar ordering of neighbours of related cells. The next state rule given by the formula $(B1D1)$ in Sect. 2.3 (Chap. 2), where a specific distribution of values in the neighbouring cells of each cell is important, establishes semantics of a type called DVN (Distribution of Values of Neighbours). Such semantics were used in the example of "two light rays" (Fig. 2.12) in Chap. 2. In this case, the similar ordering of cell neighbours in a 2-dimensional cellular structure and a behaviourally equivalent 3-dimensional one, must take place. Another type of semantics, established by the next-state rule given by the formula $(BD)$ in Sect. 2.2 (Chap. 2), where only the number of individual values in the neighbours of the cells matters, not their distribution among the neighbours, is called CVN (Counting Values of Neighbours). Semantics of this type was applied for behaviourally equivalent cellular c-e structures $V_2$ and $V_3$ shown in Fig. 4.7. The semantics of types DVN and CVN will be considered more generally in Chap. 7.
2. The location of polyhedrons, i.e. cells, in 3-dimensional Cartesian space, such as the cubes and prisms in the last two examples, depends on the coordinates in which they are located.[12] This location is similar for each polyhedron, that is, it differs only in displacement (sometimes called "translation", i.e. shift, not rotation or reflection) in space of one relative to the other. Since according to perception admitted here, each face of the polyhedron uniquely corresponds to one neighbour of the cell, the neighbours of all cells are ordered similarly, i.e. in the same succession. So, as in the case of 2-dimensional cellular c-e structures, in

[12] Notice that here the integer numbers only are used as the coordinates of cells.

the 3-dimensional ones all atoms (Sect. 2.1, Chap. 2) are isomorphic. In the last two examples, it was possible to establish isomorphisms between structures with cubic and hexagonal cells (Fig. 4.5), as well as between structures with prismatic and octagonal cells (Fig. 4.7), provided that the space coverage (tessellation) by these polyhedrons was not required. In the case of the DVN type semantics, such an isomorphism assures a one-to-one relationship between cells in 2-dimensional and 3-dimensional structures, along with the same neighbour order of each cell in both structures. Thus identical behaviour of the planar and spatial cellular c-e structures, if the behaviour is defined by the parameters $B1$, $D1$ occurring in the formulas $(B1D1)$ (Sect. 2.3, Chap. 2), specified for these isomorphic structures. In the case of the CVN type semantics, used in the example of the Game of life, the identical behaviour of the planar and spatial cellular c-e structures requires a one-to-one relationship between their cells, and only the same number of neighbours of each planar and spatial cell. But it requires application of the parameters $B$, $D$ occurring in formulas $(BD)$ (Sect. 2.2, Chap. 2), specified for these related structures. So the neighbours of each planar cell and neighbours of each related spatial cell might be ordered in a different succession.

3. We have seen that the establishment of isomorphism and similar behaviour between planar and spatial cellular c-e structures depended on their characteristics, such as the shape of polygons and polyhedrons representing cells, their arrangement in coordinate systems, the possibility or impossibility of covering by them a plane or space and the type of semantics under consideration. This problem also arises when considering the behaviour of structures in higher dimensions, in the case of manifolds of different dimensions. That is, a matter of conditions under which such structures behave identically. In particular, when the cellular c-e structures of higher dimension can be transformed into structures of lower dimension preserving behaviour. This problem is not discussed in this book.

## References

1. Bramer, M.: Logic Programming with Prolog. Springer (2005)
2. Czaja, L.: Computing in Cause-Effect, Structures. Lecture Notes in Networks and Systems, vol. 331. Springer (2022)
3. Mirkowska, G., Salwicki, A.: Algorithmic Logic. Warszawa: PWN-Polish Scientific Publishers, D. Reidel Publishing Company, A Member of the Kluwer Academic Publisher Group, Dordrecht,/Boston/Lancaster/Tokyo, (1987)
4. Hu, S.-T.: Threshold Logic. University of California Press, Berkeley and Los Angeles (1965)
5. Wilson R.J.: Introduction to Graph Theory. Longman Group Ltd. London (1979)
6. Zhang, Z.-Y.: Boolean networks. In: Dubitzky, W., Wolkenhauer, O., Cho, K.H., Yokota, H. (eds.) Encyclopedia of Systems Biology. Springer, New York (2013)

# Chapter 5
# Multi-valued Cells

So far the cells assumed binary values ***true*** and ***false***, which prompted usage of somewhat non-standard formulas $(BD)$ and $(B1D1)$ in Chap. 2 (Sects. 2.2, 2.3), defining semantics. Such formulas may be straightforwardly extended to many-valued cells, redefining the state to a function $s : \mathbb{X} \rightarrow \mathbb{Q}$ where as before, $\mathbb{X}$ denotes a universe of all nodes-cells, and $\mathbb{Q} = \{\mathbf{0}, \mathbf{1}, \mathbf{2}, \ldots, \mathbf{m}\}$, where the boldface numerals, denote values ("status") which any cell may assume. Rephrasing these formulas to many-valued cells, the restriction of the state $s$ to a set of cells $A \subseteq \mathbb{X}$ or functions of the type $A \overset{\text{into}}{\rightarrow} \mathbb{Q}$ will be used. Function $s|_A(a) \overset{\text{def}}{=} s(a)$ for $a \in A$ is such a restriction.

Before the next-state rule for many-valued cells is formulated, let us, firstly, rephrase the formula $(B1D1)$ to a new form if $\mathbb{Q} = \{\mathbf{0}, \mathbf{1}\}$. Let $a_{x,y} \notin s$ and $a_{x,y} \in s$, be written as $s(a_{x,y}) = \mathbf{0}$ and $s(a_{x,y}) = \mathbf{1}$. Consistently, $nbh(a_{x,y}) \cap s \in B1$ meaning "the set of active (in $s$) neighbours of $a_{x,y}$ belongs to family $B1$" will be now written $s|_{nbh(a_{x,y})} \in B_{\mathbf{0}\rightarrow\mathbf{1}}$ where $B_{\mathbf{0}\rightarrow\mathbf{1}}$ is a set of functions of type $nbh(a_{x,y}) \overset{\text{into}}{\rightarrow} \{\mathbf{0}, \mathbf{1}\}$ that cause change status of $a_{x,y}$ from $\mathbf{0}$ to $\mathbf{1}$. Similarly, $nbh(a_{x,y}) \cap s \in D1$ meaning "the set of active (in $s$) neighbours of $a_{x,y}$ belongs to family $D1$" will be now written $s|_{nbh(a_{x,y})} \in B_{\mathbf{1}\rightarrow\mathbf{1}}$ where $B_{\mathbf{1}\rightarrow\mathbf{1}}$ is a set of functions of type $nbh(a_{x,y}) \overset{\text{into}}{\rightarrow} \{\mathbf{0}, \mathbf{1}\}$ that cause to remain active status of $a_{x,y}$ in the state $t$ that follows $s$. The passive status of $a_{x,y}$, that is when $s(a_{x,y}) = \mathbf{0}$, remains passive, when $s|_{nbh(a_{x,y})} \notin B_{\mathbf{0}\rightarrow\mathbf{1}}$. Similarly, the active status of $a_{x,y}$, that is when $s(a_{x,y}) = \mathbf{1}$, becomes passive when $s|_{nbh(a_{x,y})} \notin B_{\mathbf{1}\rightarrow\mathbf{1}}$. Therefore, the next-state rule for a cell $a_{x,y}$ is a relation:

$$\left.\begin{array}{l} (s,t) \in [\![\, a_{x,y} \,]\!] \text{ holds if and only if } t(a_{x,y}) = \mathbf{1} \quad \text{and} \\ \\ (s(a_{x,y}) = \mathbf{0} \wedge s|_{nbh(a_{x,y})} \in B_{\mathbf{0}\rightarrow\mathbf{1}}) \vee (s(a_{x,y}) = \mathbf{1} \wedge s|_{nbh(a_{x,y})} \in B_{\mathbf{1}\rightarrow\mathbf{1}}) \end{array}\right\} (BB)$$

As before, the semantics of a cellular c-e structure $U$ is:

$$(s,t) \in [\![\, U \,]\!] \overset{\text{def}}{\Leftrightarrow} \forall a_{x,y} \in car(U) : (s,t) \in [\![\, a_{x,y} \,]\!]$$



L. Czaja, *Cellular Cause-Effect Structures*, Studies in Systems, Decision and Control 526,
https://doi.org/10.1007/978-3-031-55590-9_5

**Table 5.1** $B^{birth}_{0\to1}$–"birth" inside $a_{x,y}$ at $U_{vonNeum}$

| $nbh(a_{x,y})$ : | $a_{x-1,y}$ | $a_{x,y+1}$ | $a_{x+1,y}$ | $a_{x,y-1}$ |
|---|---|---|---|---|
| $f_1^{0\to1}$ | 1 | 1 | 1 | 0 |
| $f_2^{0\to1}$ | 1 | 1 | 0 | 1 |
| $f_3^{0\to1}$ | 1 | 0 | 1 | 1 |
| $f_4^{0\to1}$ | 0 | 1 | 1 | 1 |

**Table 5.2** $B^{surv}_{1\to1}$–"survival" inside $a_{x,y}$ at $U_{vonNeum}$

| $nbh(a_{x,y})$ : | $a_{x-1,y}$ | $a_{x,y+1}$ | $a_{x+1,y}$ | $a_{x,y-1}$ |
|---|---|---|---|---|
| $f_1^{1\to1}$ | 0 | 0 | 1 | 1 |
| $f_2^{1\to1}$ | 0 | 1 | 1 | 0 |
| $f_3^{1\to1}$ | 1 | 1 | 0 | 0 |
| $f_4^{1\to1}$ | 0 | 1 | 0 | 1 |
| $f_5^{1\to1}$ | 1 | 0 | 1 | 0 |
| $f_6^{1\to1}$ | 1 | 0 | 0 | 1 |
| $f_7^{1\to1}$ | 1 | 1 | 1 | 0 |
| $f_8^{1\to1}$ | 1 | 1 | 0 | 1 |
| $f_9^{1\to1}$ | 1 | 0 | 1 | 1 |
| $f_{10}^{1\to1}$ | 0 | 1 | 1 | 1 |

The parameters $B_{0\to1}$, $B_{1\to1}$ are sets of functions $f_k^{\mathbf{j}\to\mathbf{i}} : nbh(a_{x,y}) \overset{\text{into}}{\to} \{\mathbf{0}, \mathbf{1}\}$ $(\mathbf{j}, \mathbf{i} \in \{\mathbf{0}, \mathbf{1}\})$ called patterns determining change (or not) of the status of $a_{x,y}$. This takes place when the restricted state $s|_{nbh(a_{x,y})}$ matches a pattern belonging to $B_{0\to1}$ or $B_{1\to1}$. The patterns specify a concrete task the cellular c-e structure is to perform. [1]

For instance, the patterns for the von Neumann type of "Game of life" (Fig. 2.1) are functions in sets

$$B^{birth}_{0\to1} = \{f_1^{0\to1}, f_2^{0\to1}, f_3^{0\to1}, f_4^{0\to1}\}$$

("birth in the cell $a_{x,y}$") and

$$B^{surv}_{1\to1} = \{f_1^{1\to1}, f_2^{1\to1}, f_3^{1\to1}, f_4^{1\to1}, f_5^{1\to1}, f_6^{1\to1}, f_7^{1\to1}, f_8^{1\to1}, f_9^{1\to1}, f_{10}^{1\to1}\}$$

("survival in the cell $a_{x,y}$"), listed in the Tables 5.1 and 5.2[2]:

Notice that the same order and orientation of each cell's neighbours should be assumed in this formulation of semantics defined by functions $f_k^{0\to1}$ and $f_k^{1\to1}$ listed in tables $B^{birth}_{0\to1}$ and $B^{surv}_{1\to1}$. Thus to treat the neighbourhoods of cells as vectors (not

[1] The tables $B_{\mathbf{j}\to\mathbf{i}}$ of patterns $f_k^{\mathbf{j}\to\mathbf{i}}$ here take place role of global transition functions of many arguments, in cellular automata.

[2] Note that all functions used here are of a single argument and are the same for every cell $a_{x,y} \in \mathbb{X}$.

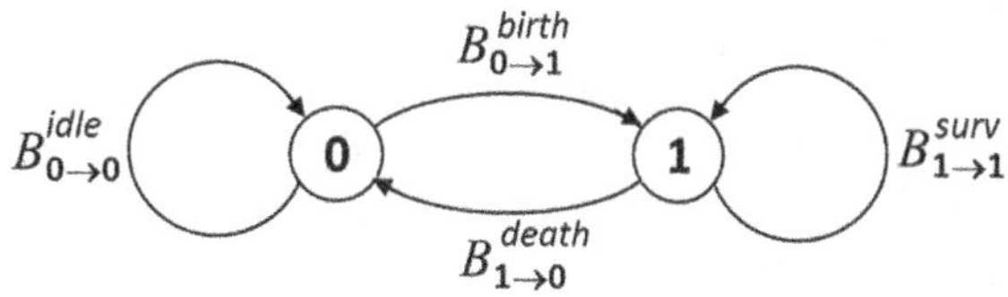


**Fig. 5.1** Changes of cell's $a_{x,y}$ status in the von Neumann Game of life. This is a shorthand of this multigraph in which each edge is labelled by functions (patterns) belonging to sets (parameters) $B_{j\to i}$. Let $B^{death}_{1\to 0} = \overline{B^{surv}_{1\to 1}}$ and $B^{idle}_{0\to 0} = \overline{B^{birth}_{0\to 1}}$, since, if e.g. $f \in B^{death}_{1\to 0} \cap B^{surv}_{1\to 1}$ then in this multigraph, the pattern $f$ would label different edges $\mathbf{1} \to \mathbf{0}$ and $\mathbf{1} \to \mathbf{1}$. The symbols $\overline{B^{surv}_{1\to 1}}$, $\overline{B^{birth}_{0\to 1}}$ denote complements of $B^{surv}_{1\to 1}$ , $B^{birth}_{0\to 1}$ to the set of all functions of the type $nbh(a_{x,y}) \overset{\text{into}}{\to} \{\mathbf{0}, \mathbf{1}\}$. The edges labelled by $B^{death}_{1\to 0}$ and $B^{idle}_{0\to 0}$ are being traversed when $s|_{nbh(a_{x,y})} \notin B^{surv}_{1\to 1}$ and $s|_{nbh(a_{x,y})} \notin B^{birth}_{0\to 1}$

just as sets). For semantics defined in terms of tables (let call them semantic tables), that will be done since now, the atoms of cellular c-e structures are assumed equivalent relative-to the shift transformation. So far this played only illustrative part for examples.

Behaviour of the cell $a_{x,y}$ in the von Neumann Game of life may be presented as the automaton in Fig. 5.1.

### Remark

Perceiving a cellular c-e structure as a set of cells changing their status according to their local rules, prompts perception of each cell $a_{x,y}$ as an automaton changing its state depending on its neighbourhood associated with an input symbol $B$. In the setting put forth here, the input symbols are parameters $B$, each being a set of patterns, i.e. functions of the type $nbh(a_{x,y}) \to \mathbb{Q}$. So, the graph of automaton in Fig. 5.1 is a multigraph, since every edge, in fact represents a bunch of edges labelled with the patterns: transit between a pair of states may progress along diverse paths (like in quantum phenomena). The automaton makes consistency check of its current state with the patterns and changes its state to a state determined by a certain pattern in $B$–if its current state matches such pattern there.

## *Example (More Light Rays)*

As another example of using parameters $B$, let us mention the "two light rays" cellular c-e structure (Fig. 2.12), but rephrasing its parameters to be sets of functions $B^{red}_{0\to 1} = \{f^{0\to 1}_1, f^{0\to 1}_2, f^{0\to 1}_3, f^{0\to 1}_4\}$ assuming values: $\mathbf{0}$ (white), $\mathbf{1}$ (red), $\omega$ (yellow)

**Table 5.3** Defining set $B^{red}_{\mathbf{0}\to\mathbf{1}}$ of functions $f_k^{\mathbf{0}\to\mathbf{1}}$ with domain $nbh(a_{x,y})$ filling up the top row and with values in the rows underneath. The table defines formation by cellular c-e structure $U_{red}$ two red light rays that emerge from the yellow cell $a_{0,0}$ in opposite directions shown in Fig. 5.2 if initial state $s$ is $s(a_{0,0}) = \omega$ and $s(a_{x,y}) = \mathbf{0}$ for remaining $x$, $y$

| $nbh(a_{x,y})$ : | $a_{x-1,y-1}$ | $a_{x-1,y+1}$ | $a_{x,y+2}$ | $a_{x+1,y+1}$ | $a_{x+1,y-1}$ | $a_{x,y-2}$ |
|---|---|---|---|---|---|---|
| red ray $f_1^{\mathbf{0}\to\mathbf{1}}$ | $\mathbf{0}$ | $\mathbf{1}$ | $\mathbf{0}$ | $\mathbf{0}$ | $\mathbf{0}$ | $\mathbf{0}$ |
| red ray $f_2^{\mathbf{0}\to\mathbf{1}}$ | $\mathbf{0}$ | $\mathbf{0}$ | $\mathbf{0}$ | $\mathbf{0}$ | $\mathbf{1}$ | $\mathbf{0}$ |
| red ray $f_3^{\mathbf{0}\to\mathbf{1}}$ | $\mathbf{0}$ | $\omega$ | $\mathbf{0}$ | $\mathbf{0}$ | $\mathbf{0}$ | $\mathbf{0}$ |
| red ray $f_4^{\mathbf{0}\to\mathbf{1}}$ | $\mathbf{0}$ | $\mathbf{0}$ | $\mathbf{0}$ | $\mathbf{0}$ | $\omega$ | $\mathbf{0}$ |

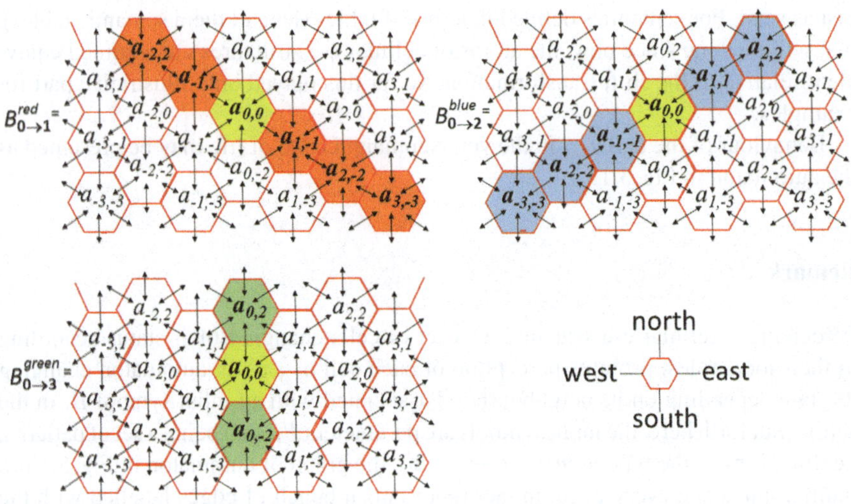


**Fig. 5.2** Image of three pieces of cellular c-e structures $U_{red}$, $U_{blue}$, $U_{green}$ producing red, blue and green light rays propagating from the yellow cell $a_{0,0}$. A white cell in respective structure turns to: red when its southeast or northwest neighbour is yellow or red, blue when its southwest or northeast neighbour is yellow or blue, green when its south or north neighbour is yellow or green. Otherwise, the white cell remains white, The yellow, red, blue and green cells never change value (colour)

defined in Table 5.3 that, along with remaining sets $B^{red}_{\mathbf{j}\to\mathbf{i}}$, make picture of two red light rays propagating from the yellow cell $a_{0,0}$ in Fig. 5.2.[3]

Remaining sets $B^{red}_{\mathbf{j}\to\mathbf{i}}$ are: $B^{red}_{\mathbf{1}\to\mathbf{1}} = B^{red}_{\omega\to\omega} = \{\mathbf{0}, \mathbf{1}, \omega\}^{nbh(a_{x,y})}$ thus if $a_{x,y}$ is red or yellow then it never changes colour, and $B^{red}_{\mathbf{0}\to\mathbf{0}} = \overline{B^{red}_{\mathbf{0}\to\mathbf{1}}}$ thus if $a_{x,y}$ is white then it remains white when $s|_{nbh(a_{x,y})} \notin B^{red}_{\mathbf{0}\to\mathbf{1}}$ and $B^{red}_{\mathbf{0}\to\omega} = B^{red}_{\mathbf{1}\to\mathbf{0}} = B^{red}_{\mathbf{1}\to\omega} = B^{red}_{\omega\to\mathbf{0}} = B^{red}_{\omega\to\mathbf{1}} = \varnothing$.

[3] Recall that hexagonal cells covering the plane, have both coordinates $x$, $y$ of the same parity.

The relation $[\![ a_{x,y} ]\!]$ that determines behaviour of a cell $a_{x,y}$ may be written as: $(s,t) \in [\![ a_{x,y} ]\!]$ holds if and only if certain of the following formulas holds:

$$\left[\bigvee_{\mathbf{j}\in\{\mathbf{0},\mathbf{1},\omega\}} s(a_{x,y}) = \mathbf{j} \wedge s|_{nbh(a_{x,y})} \in B^{red}_{\mathbf{j}\to\mathbf{0}}\right] \wedge t(a_{x,y}) = \mathbf{0}$$

$$\left[\bigvee_{\mathbf{j}\in\{\mathbf{0},\mathbf{1},\omega\}} s(a_{x,y}) = \mathbf{j} \wedge s|_{nbh(a_{x,y})} \in B^{red}_{\mathbf{j}\to\mathbf{1}}\right] \wedge t(a_{x,y}) = \mathbf{1}$$

$$\left[\bigvee_{\mathbf{j}\in\{\mathbf{0},\mathbf{1},\omega\}} s(a_{x,y}) = \mathbf{j} \wedge s|_{nbh(a_{x,y})} \in B^{red}_{\mathbf{j}\to\omega}\right] \wedge t(a_{x,y}) = \omega,$$

where $\bigvee_{\mathbf{j}\in\{\mathbf{0},\mathbf{1},\omega\}}$ stands for repeated disjunction. In words: $(s,t) \in [\![ a_{x,y} ]\!]$ holds iff $s(a_{x,y}) = \mathbf{j}$, $t(a_{x,y}) = \mathbf{i}$ and neighbours of $a_{x,y}$ in the state $s$ have values occurring in certain row of the set $B^{red}_{\mathbf{j}\to\mathbf{i}}$ ($\mathbf{j},\mathbf{i} = \mathbf{0},\mathbf{1},\omega$). To put it shorter: $(s,t) \in [\![ a_{x,y} ]\!]$ holds iff certain of these three formulas is true. Note that $s|_{nbh(a_{x,y})} \in B^{red}_{\mathbf{1}\to\mathbf{1}}$ and $s|_{nbh(a_{x,y})} \in B^{red}_{\omega\to\omega}$ are always true. Concisely:

$$\bigvee_{\mathbf{i}\in\{\mathbf{0},\mathbf{1},\omega\}} \left(\left[\bigvee_{\mathbf{j}\in\{\mathbf{0},\mathbf{1},\omega\}} s(a_{x,y}) = \mathbf{j} \wedge s|_{nbh(a_{x,y})} \in B^{red}_{\mathbf{j}\to\mathbf{i}}\right] \wedge t(a_{x,y}) = \mathbf{i}\right)$$

Similarly, the set $B^{blue}_{\mathbf{0}\to\mathbf{2}} = \{f_1^{\mathbf{0}\to\mathbf{2}}, f_2^{\mathbf{0}\to\mathbf{2}}, f_3^{\mathbf{0}\to\mathbf{2}}, f_4^{\mathbf{0}\to\mathbf{2}}\}$ whose functions assume values $\mathbf{0}$ (white), $\mathbf{2}$ (blue), $\omega$ (yellow) defined in Table 5.4 that, along with remaining $B^{blue}_{\mathbf{j}\to\mathbf{i}}$, make picture of two blue light rays propagating from the yellow cell $a_{0,0}$ in cellular c-e structure $U_{blue}$ shown in Fig. 2.33. Thus $(s,t) \in [\![ a_{x,y} ]\!]$ holds if and only if

$$\bigvee_{\mathbf{i}\in\{\mathbf{0},\mathbf{2},\omega\}} \left(\left[\bigvee_{\mathbf{j}\in\{\mathbf{0},\mathbf{2},\omega\}} s(a_{x,y}) = \mathbf{j} \wedge s|_{nbh(a_{x,y})} \in B^{blue}_{\mathbf{j}\to\mathbf{i}}\right] \wedge t(a_{x,y}) = \mathbf{i}\right)$$

Remaining sets $B^{blue}_{\mathbf{j}\to\mathbf{i}}$ are: $B^{blue}_{\mathbf{2}\to\mathbf{2}} = B^{blue}_{\omega\to\omega} = \{\mathbf{0},\mathbf{2},\omega\}^{nbh(a_{x,y})}$ thus if $a_{x,y}$ is blue or yellow then it never changes colour, and $B^{blue}_{\mathbf{0}\to\mathbf{0}} = \overline{B^{blue}_{\mathbf{0}\to\mathbf{2}}}$ thus if $a_{x,y}$ is white then it remains white when $s|_{nbh(a_{x,y})} \notin B^{blue}_{\mathbf{0}\to\mathbf{2}}$ and $B^{blue}_{\mathbf{0}\to\omega} = B^{blue}_{\mathbf{2}\to\mathbf{0}} = B^{blue}_{\mathbf{2}\to\omega} = B^{blue}_{\omega\to\mathbf{0}} = B^{blue}_{\omega\to\mathbf{2}} = \varnothing$.

And finally, the set $B^{green}_{\mathbf{0}\to\mathbf{3}} = \{f_1^{\mathbf{0}\to\mathbf{3}}, f_2^{\mathbf{0}\to\mathbf{3}}, f_3^{\mathbf{0}\to\mathbf{3}}, f_4^{\mathbf{0}\to\mathbf{3}}\}$ whose functions assume values $\mathbf{0}$ (white), $\mathbf{3}$ (green), $\omega$ (yellow) defined in Table 5.2, that along with remaining $B^{green}_{\mathbf{j}\to\mathbf{i}}$, make picture of two green light rays propagating from the yellow cell $a_{0,0}$ in cellular c-e structure $U_{green}$ shown in Fig. 5.2. Thus $(s,t) \in [\![ a_{x,y} ]\!]$ holds if and only if

**Table 5.4** Defining set $B^{blue}_{0\to 2}$ of functions $f_k^{0\to 2}$ with domain $nbh(a_{x,y})$ filling up the top row and with values in the rows underneath. The table defines formation by cellular c-e structure $U_{blue}$ two blue light rays that emerge from the yellow cell $a_{0,0}$ in opposite directions shown in Fig. 5.2 if initial state $s$ is $s(a_{0,0}) = \omega$ and $s(a_{x,y}) = \mathbf{0}$ for remaining $x$, $y$

| $nbh(a_{x,y})$ : | $a_{x-1,y-1}$ | $a_{x-1,y+1}$ | $a_{x,y+2}$ | $a_{x+1,y+1}$ | $a_{x+1,y-1}$ | $a_{x,y-2}$ |
|---|---|---|---|---|---|---|
| blue ray $f_1^{0\to 2}$ | **2** | **0** | **0** | **0** | **0** | **0** |
| blue ray $f_2^{0\to 2}$ | **0** | **0** | **0** | **2** | **0** | **0** |
| blue ray $f_3^{0\to 2}$ | $\omega$ | **0** | **0** | **0** | **0** | **0** |
| blue ray $f_4^{0\to 2}$ | **0** | **0** | **0** | $\omega$ | **0** | **0** |

**Table 5.5** Defining set $B^{green}_{0\to 3}$ of functions $f_k^{0\to 3}$ with domain $nbh(a_{x,y})$ filling up the top row and with values in the rows underneath. The table defines formation by cellular c-e structure $U_{green}$ two green light rays that emerge from the yellow cell $a_{0,0}$ in opposite directions shown in Fig. 5.2 if initial state $s$ is $s(a_{0,0}) = \omega$ and $s(a_{x,y}) = \mathbf{0}$ for remaining $x$, $y$

| $nbh(a_{x,y})$ : | $a_{x-1,y-1}$ | $a_{x-1,y+1}$ | $a_{x,y+2}$ | $a_{x+1,y+1}$ | $a_{x+1,y-1}$ | $a_{x,y-2}$ |
|---|---|---|---|---|---|---|
| green ray $f_1^{0\to 3}$ | **0** | **0** | **3** | **0** | **0** | **0** |
| green ray $f_2^{0\to 3}$ | **0** | **0** | **0** | **0** | **0** | **3** |
| green ray $f_3^{0\to 3}$ | **0** | **0** | $\omega$ | **0** | **0** | **0** |
| green ray $f_4^{0\to 3}$ | **0** | **0** | **0** | **0** | **0** | $\omega$ |

$$\bigvee_{\mathbf{i}\in\{\mathbf{0},\mathbf{3},\omega\}} \left( \left[ \bigvee_{\mathbf{j}\in\{\mathbf{0},\mathbf{3},\omega\}} s(a_{x,y}) = \mathbf{j} \wedge s|_{nbh(a_{x,y})} \in B^{green}_{\mathbf{j}\to\mathbf{i}} \right] \wedge t(a_{x,y}) = \mathbf{i} \right)$$

Remaining sets $B^{green}_{\mathbf{j}\to\mathbf{i}}$ are: $B^{green}_{\mathbf{3}\to\mathbf{3}} = B^{green}_{\omega\to\omega} = \{\mathbf{0}, \mathbf{3}, \omega\}^{nbh(a_{x,y})}$ thus if $a_{x,y}$ is green or yellow then it never changes colour, and $B^{green}_{\mathbf{0}\to\mathbf{0}} = \overline{B^{green}_{\mathbf{0}\to\mathbf{3}}}$ thus if $a_{x,y}$ is white then it remains white when $s|_{nbh(a_{x,y})} \notin B^{green}_{\mathbf{0}\to\mathbf{3}}$ and $B^{green}_{\mathbf{0}\to\omega} = B^{green}_{\mathbf{3}\to\mathbf{0}} = B^{green}_{\mathbf{3}\to\omega} = B^{green}_{\omega\to\mathbf{0}} = B^{green}_{\omega\to\mathbf{3}} = \varnothing$

It should be noticed that the cellular c-e structures $U_{red}$, $U_{blue}$, $U_{green}$ with the same carriers and neighbourhoods, are structurally identical, but semantically different. Their behaviour is determined by next-state rules based on different semantic Tables 5.3, 5.4, 5.5. Each structure starts activity from the cell $a_{0,0}$ of value $\omega$ (remaining cells of value **0**) and fills up cells with values **1** (red), **2** (blue) and **3** (green) along directions depicted in Fig. 5.2.

Semantics of cellular c-e structure $U \in \{U_{red}, U_{blue}, U_{green}\}$ is a binary relation:

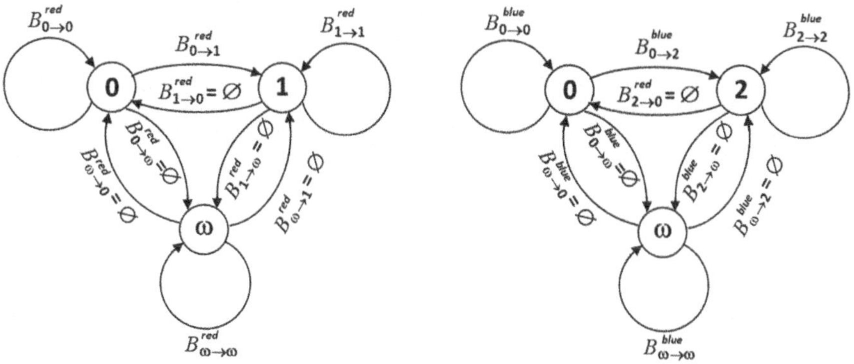


**Fig. 5.3** Changes of cell $a_{x,y}$ status during activity of the cellular c-e structures generating red and blue light rays. The edges labelled by empty sets $B$ are never traversed thus may be deleted for this task

**Fig. 5.4** Changes of cell $a_{x,y}$ status during activity of the cellular c-e structures generating green light rays

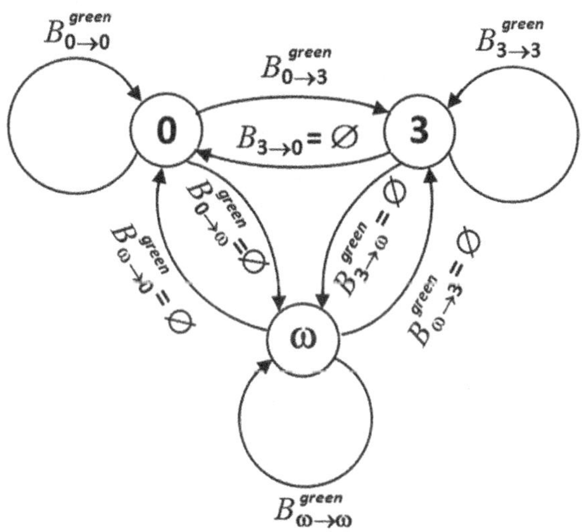


$$(s, t) \in [\![ U ]\!] \overset{\text{def}}{\Leftrightarrow} \forall a_{x,y} \in car(U) : (s, t) \in [\![ a_{x,y} ]\!]$$

Behaviour of the cell $a_{x,y}$ of cellular c-e structures generating these coloured light rays may be presented as the automata in Figs. 5.3, 5.4.

**End of example "more light rays".**

After these examples explaining the formulas of the next-state rules for cellular c-e structures with two and three valued cells, the formulation of semantics in general case, when $\mathbb{Q} = \{\mathbf{0}, \mathbf{1}, \mathbf{2}, \ldots, \mathbf{m}\}$ is straightforward. The local next-state rule for a cell $a$ changing state $s$ into $t$ is the following:

$(s, t) \in [\![ a ]\!]$ holds if and only if a certain of the following $m + 1$ formulas holds:

$$\left[\bigvee_{\mathbf{j} \in \mathbf{Q}} s(a) = \mathbf{j} \wedge s|_{nbh(a)} \in B_{\mathbf{j} \to \mathbf{0}}\right] \wedge t(a) = \mathbf{0}$$

$$\left[\bigvee_{\mathbf{j} \in \mathbf{Q}} s(a) = \mathbf{j} \wedge s|_{nbh(a)} \in B_{\mathbf{j} \to \mathbf{1}}\right] \wedge t(a) = \mathbf{1}$$

$$\left[\bigvee_{\mathbf{j} \in \mathbf{Q}} s(a) = \mathbf{j} \wedge s|_{nbh(a)} \in B_{\mathbf{j} \to \mathbf{2}}\right] \wedge t(a) = \mathbf{2}$$

..............................................

$$\left[\bigvee_{\mathbf{j} \in \mathbf{Q}} s(a) = \mathbf{j} \wedge s|_{nbh(a)} \in B_{\mathbf{j} \to \mathbf{m}}\right] \wedge t(a) = \mathbf{m}$$

In the concise form, the next-state rule $[\![ a ]\!]$ for the cell $a$ is:

$$(s, t) \in [\![ a ]\!] \overset{\text{def}}{\Leftrightarrow} \bigvee_{\mathbf{i} \in \mathbf{Q}} \left( \left[\bigvee_{\mathbf{j} \in \mathbf{Q}} s(a) = \mathbf{j} \wedge s|_{nbh(a)} \in B_{\mathbf{j} \to \mathbf{i}}\right] \wedge t(a) = \mathbf{i} \right)$$

In words: the cell $a$ changes its value $\mathbf{j}$ in the state $s$ to value $\mathbf{i}$ in the next-state $t$ iff $s|_{nbh(a)} \in B_{\mathbf{j} \to \mathbf{i}}$ holds.[4]

Semantics of the cellular c-e structure $U$ in general case, is defined as the following binary relation between state $s$ and its successor $t$:

$$(s, t) \in [\![ U ]\!] \overset{\text{def}}{\Leftrightarrow} \forall a \in car(U) : (s, t) \in [\![ a ]\!]$$

Note that dimension of cellular c-e structures is here immaterial, so we write "$a$" instead of "$a_{x,y,\ldots,z}$". Remember that in contrast to what is adopted in cellular automata literature, here the cells do not belong to their neighbourhood. So, if each cell has $n$ neighbours and may assume $m + 1$ values (as in the above set up), the number of all possible local next-state rules amounts to $(m + 1)^{(m+1)^{(n+1)}}$.

It may be expedient to illustrate this definitional formula for semantics by an example, sometimes exploited in the cellular automata studies. The purpose is closer explanation of a role the parameters $B_{\mathbf{j} \to \mathbf{i}}$ play in devising a cellular c-e structure. To keep this simple, let confine ourselves to one dimensional case with two valued cells.

---

[4] Strictly: states $s$ and $t$ are interrelated by means of relation $[\![ a ]\!]$. The phrase "changes its value" refers to the intentional role of state $t$ as the next to $s$.

## *Example (Motion of Vehicles on Straight Road)*

A number of vehicles are heading on straight (infinite) road rightwards one after one, without outrunning one another. Figure 5.5 shows a 1-dimensional cellular c-e structure $U$ with values **0**, **1** of cells. It behaves in accordance with semantics $[\![ U ]\!]$ imposing traffic regulation and is constructed as follows.

In accordance with the above schema of the local next-state rule $(s, t) \in [\![ a_x ]\!]$ we get that a cell $a_x$ passes its status to **0** when either its current (in the state $s$) status is **0** and the set $B_{0\to0}$ contains a function identical with $s|_{nbh(a_x)}$ or its current status is **1** and the set $B_{1\to0}$ contains a function identical with $s|_{nbh(a_x)}$. Similarly, $a_x$ passes its status to **1** when its current status is **0** and the set $B_{0\to1}$ contains a function identical with $s|_{nbh(a_x)}$ or its current status is **1** and the set $B_{1\to1}$ contains a function identical with $s|_{nbh(a_x)}$. These two cases are expressed by formulas:

$$\big[(s(a_x) = \mathbf{0} \wedge s|_{nbh(a_x)} \in B_{0\to0}) \vee (s(a_x) = \mathbf{1} \wedge s|_{nbh(a_x)} \in B_{1\to0})\big] \wedge t(a_x) = \mathbf{0}$$

$$\big[(s(a_x) = \mathbf{0} \wedge s|_{nbh(a_x)} \in B_{0\to1}) \vee (s(a_x) = \mathbf{1} \wedge s|_{nbh(a_x)} \in B_{1\to1})\big] \wedge t(a_x) = \mathbf{1}$$

and $(s, t) \in [\![ a_x ]\!]$ is defined as their disjunction. The only inventive step in this construction is to specify one-argument functions ("patterns") making sets $B_{0\to0}$, $B_{1\to0}$, $B_{0\to1}$, $B_{1\to1}$, thus to express intended ongoings of vehicles. In this task it is immediate for a normal traffic, so the tables of patterns are the following, where **0** and **1** mean absence and presence of a vehicle at a cell:

$B_{0\to0} =$

| $nbh(a_x)$ : | $a_{x-1}$ | $a_{x+1}$ |
|---|---|---|
| $f_1^{0\to0}$ | **0** | **0** |
| $f_2^{0\to0}$ | **0** | **1** |

$B_{0\to1} =$

| $nbh(a_x)$ : | $a_{x-1}$ | $a_{x+1}$ |
|---|---|---|
| $f_1^{0\to1}$ | **1** | **0** |
| $f_2^{0\to1}$ | **1** | **1** |

$B_{1\to0} =$

| $nbh(a_x)$ : | $a_{x-1}$ | $a_{x+1}$ |
|---|---|---|
| $f_1^{1\to0}$ | **0** | **0** |
| $f_2^{1\to0}$ | **1** | **0** |

$B_{1\to1} =$

| $nbh(a_x)$ : | $a_{x-1}$ | $a_{x+1}$ |
|---|---|---|
| $f_1^{1\to1}$ | **0** | **1** |
| $f_2^{1\to1}$ | **1** | **1** |

Notice that the number of vehicles on the road is constant in this solution: no vehicles enter the road from outside or depart it (Fig. 5.5).

**End of example.**

A modification of the traffic on the straight road, into the circular road traffic, like on a roundabout, is obvious. To construct such cellular c-e structure, a segment of the straight road is cut-out and the auxiliary bidirectional arrow linking the segment's ends is added to it–see Fig. 4.1 in Chap. 4. The tables of patterns in the sets $B_{j\to i}$ are as for the straight road traffic with the only provision that $a_{x-1}$, $a_{x+1}$ are changed

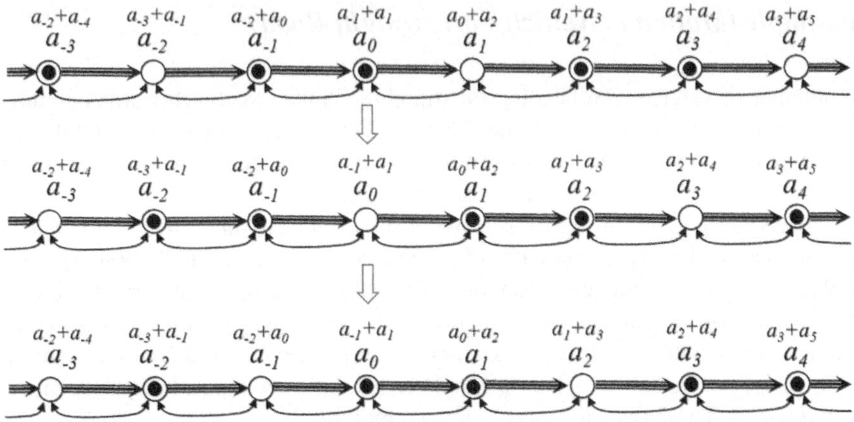


**Fig. 5.5** Three successive stages of evolving situation on the road, generating states $s1$, $s2$, $s3$ such that $s1(a_i) = 1$, for $i = -3, -1, 0, 2, 3$and $s1(a_i) = 0$, for $i = -2, 1, 4$ $s2(a_i) = 1$, for $i = -2, -1, 1, 2, 4$and $s2(a_i) = 0$, for $i = -3, 0, 3$ $s3(a_i) = 1$, for $i = -2, 0, 1, 3, 4$and $s3(a_i) = 0$, for $i = -3, -1, 2$ Combined arrows: $\Rrightarrow$ are road sections, bidirectional arrows link cells with their neighbours, tokens are vehicles

to $a_{x\ominus 1}$, $a_{x\oplus 1}$ where $\ominus$ and $\oplus$ mean subtraction and addition of integers modulo the length of the segment (see description regarding Fig. 4.1).

### Remark

The issues considered here lead to several notes related to designing cellular c-e structures. Having fixed a shape of such structure's base (topological manifold like flat or curved surface or other space), a shape of cells on it and established their neighbourhoods, one has to devise sets $B$ (parameters) of 1-argument functions called patterns that assign a fixed (independent of current state) value to each neighbour of a cell. The intended change of the cell's value takes place according to whether the current state of its neighbours, matches the patterns. The parameters $B$, presented here by tables of patterns, implement a concrete task the cellular c-e structure is to perform, starting from a given initial state. They determine next-state rule for the cellular c-e structure, so its semantics. Much attention has been given to forming bases the cellular c-e structures lie on, to the geometrical shape of cells and to possibility of tessellation of bases by the cells. These considerations exploited some topological facts and algebraic features of cause-effect structures. In particular, commutativity, associativity and independence of summation of c-e structures, and cutting them off to polygons or polyhedra, and "bending" such fragments by bringing together (by means of addition) their opposite sides. In this way e.g. cylinders and toruses as bases for the cellular c-e structures have been obtained.

Some peculiarities of construction of bases for cellular c-e structures and shape of cells should be noticed.[5] Two-dimensional flatten bases like those in Figs. 2.1, 2.6, 2.9 (Chap. 2), or curved like in Figs. 3.4, 3.10 (Chap. 3), imitate trusses (or garden lattices) on which cells are placed. Contrary to what is adopted in the cellular automata, where the cells reside at truss nodes, here the cells are imitated by polygonal tiles, whose vertices (corners) are the truss nodes. The construction is outfitted with Cartesian coordinate system in such way that each tile is identified by integer coordinates of its certain inner point. This arrangement enables construction of any cellular c-e structure by summing up its building blocks called atoms, each being a cell-tile surrounded by its nearest neighbouring cells-tiles. As result of this construction, any tile can or cannot share its sides with neighbouring tiles, depending on a kind of polygons making the tiles. Although only triangular, rectangular and hexagonal tiles are of this kind, thus capable of covering the base without gaps and overlaps, i.e. making tessellation of the base, any polygonal tiles (all of the same kind) may be cells of two-dimensional cellular c-e structures. The capability of base tessellation is not an immanent feature of cellular c-e structures. It is absent in their definition of structure and semantics. However may be expedient for some applications. Similar notes concern cellular c-e structures of dimension 3 (or greater). The spatial bases imitate 3-dimensional trusses filled with polyhedral cells, some solid figures whose vertices are the truss nodes, as illustrated in Figs. 4.4, 4.5, 4.6, 4.7 (Chap. 4). In this case the role of polygon sides play polyhedron faces, and whether a 3-dimensional truss can or cannot be covered by identical solid figures without gaps and overlaps, depends on what kind of polyhedron the solid figure is. For instance, cannot be tessellated by tetrahedrons, but can by cubes or hexagonal prisms (Figs. 4.4, 4.7). Of course, here the possibility of tessellation means sharing faces by neighbouring cells.

[5] In writings on cellular automata, what is called here "base", is usually named "domain"

# Chapter 6
# Combining Structures into Cellular

The crucial feature of cause-effect structures is capability of their combining. Which means that addition and multiplication yields correct c-e structure. So the standard c-e structures make an algebraic calculus, the quasi semiring. Contrarily to their cellular version, different nodes of general c-e structure may have different number of predecessors (causes) and successors (effects), as well as the next-state rule is the same for all c-e structures. This is not so with the cellular ones. So far, combining atomic c-e structures—their summation—has been applied for creation of complete networks being skeleton constructs for cellular c-e structures, as exemplify Figs. 2.5, 2.6, 2.8, 2.9 in Chap. 2. However not always summation of atoms yields correct cellular c-e structures, as show Figs. 3.11, 3.12 in Chap. 3. All the more this concerns complete cellular c-e structures, the problem considered in the present and the next chapter. The correctness of a summation result depends on certain properties of components that are being summed up, especially on the distribution of neighbours around identical cells shared by the components. The result should enjoy properties inherent for the cellular automata such as similarity of all atoms in effect of their shift transformation, called ***equivalence relative-to the shift***, thus also equinumerous neighbourhoods of all cells. In combining cellular c-e structures, the positional (set theoretic) relationships between neighbourhoods of cells in the components have to be taken into regard, like disjointness (or not) or containment of neighbourhoods of sharing cells, equality of carriers, etc. Also geometric shape and arrangement of cells, if their common sides or faces means their neighbouring. As mentioned in the beginning of Chap. 2, different cellular c-e structures have various rules of the next-state. This concerns also components and result of their combining: semantics of the result is being devised to implement a modelled task.

Before entering these issues, let us consider summation of two cellular c-e structures, each being a structural representation of the von Neumann type Game of life. They are placed on the same planar Cartesian coordinate system but the neighbourhoods of identically located cells in both components of addition are differently located. Due to the idempotence of addition, the cells of identical coordinates in both components, in the result appear as one cell, surrounded by the union of neighbourhoods of these identically located cells. This is depicted in Fig. 6.1.

 
L. Czaja, *Cellular Cause-Effect Structures*, Studies in Systems, Decision and Control 526,
https://doi.org/10.1007/978-3-031-55590-9_6

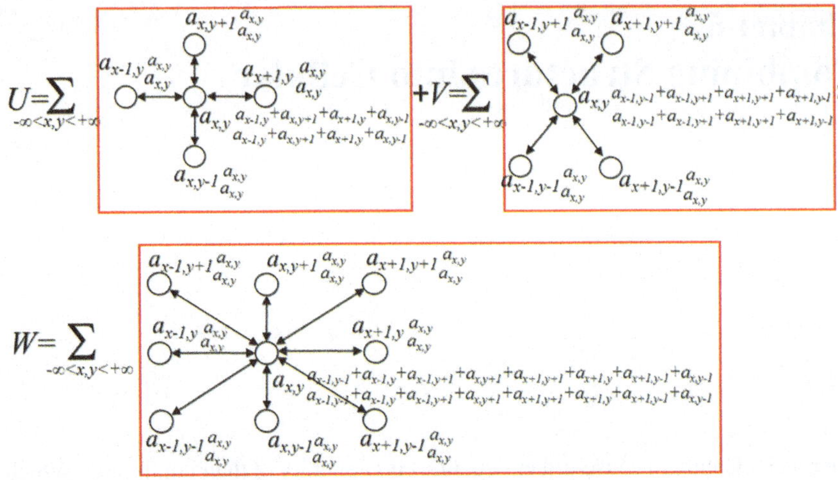


**Fig. 6.1** Summation of two cellular c-e structures Game of life of the von Neumann type (Fig. 2.4 in Chap. 2), yields the cellular c-e structure of the Moore type (Fig. 2.5)

In the Game of life, behaviour of both components $U$, $V$ of summation depends only on a number of the active neighbours of cells, not on location of active neighbours in the neighbourhoods. Each cell in the components has four neighbours, so the next-state rule for $U$ and $V$ is the same. But each cell shared by both components may behave differently or identically in individual components in the same state $s$ : $car(U) = car(V) = car(W) \to \{\mathbf{0}, \mathbf{1}\}$. One may immediately conclude behaviour of the result $W$ of summation, because the neighbourhoods of cells common to $U$ and $V$ are disjoint. So for any cell $a_{x,y}$:

$$\underset{U}{nbh}(a_{x,y}) \cap \underset{V}{nbh}(a_{x,y}) = \varnothing \quad \text{and} \quad \underset{U}{nbh}(a_{x,y}) \cup \underset{V}{nbh}(a_{x,y}) = \underset{W}{nbh}(a_{x,y})$$

$$\left|\underset{U}{nbh}(a_{x,y}) \cup \underset{V}{nbh}(a_{x,y})\right| = \left|\underset{U}{nbh}(a_{x,y})\right| + \left|\underset{V}{nbh}(a_{x,y})\right| = \left|\underset{W}{nbh}(a_{x,y})\right| = 8$$

and in any state $s$:

$$\left|\underset{U}{nbh}(a_{x,y}) \cap s\right| + \left|\underset{V}{nbh}(a_{x,y}) \cap s\right| = \left|\underset{W}{nbh}(a_{x,y}) \cap s\right|$$

which tells how the cell $a_{x,y}$ in the result $W$ would change (or not) its status (live/dead) depending on status of neighbours of this cell in components $U$ and $V$. ($\underset{U}{nbh}(a_{x,y})$ means that this neighbourhood is in the cellular c-e structure $U$, etc.)

In order to present summation of cellular c-e structures more visible graphically, Fig. 6.2 shows fragments of structures $U$, $V$, $W$ from Fig. 6.1 around the cell $a_{0,0}$ as graphs embedded into the Cartesian coordinate system.

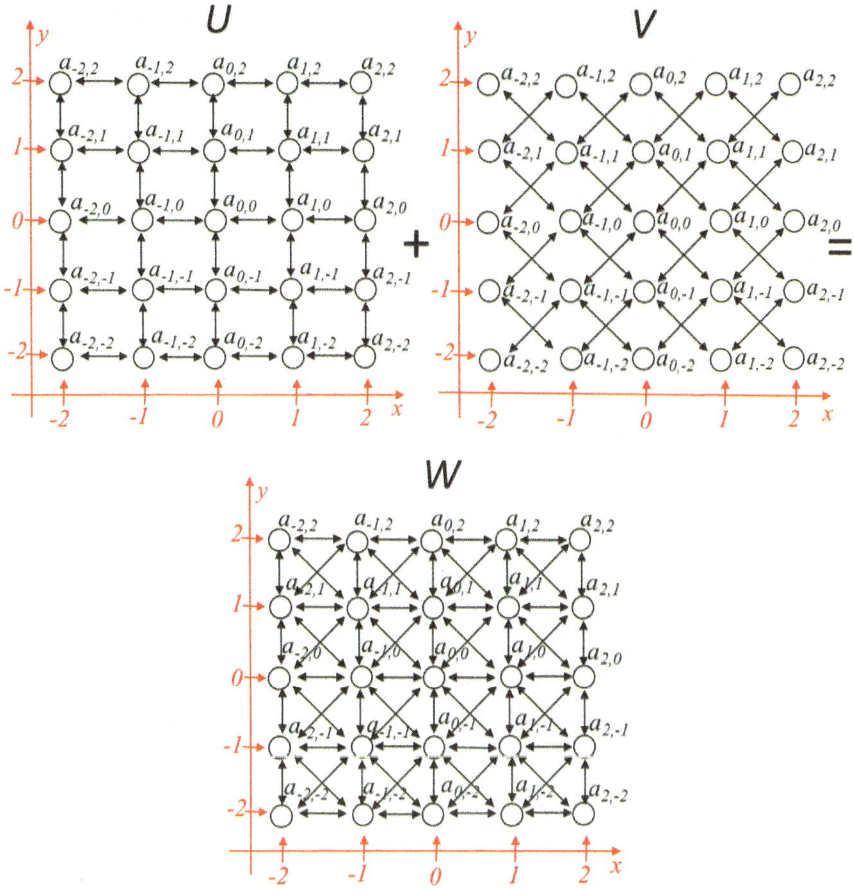


**Fig. 6.2** Fragments of cellular c-e structures $U$, $V$, $W$; the outcome of summation $W = U + V$. Distribution of cells in $U$ and $V$ on the base—as in Fig. 2.1, and in $W$—as in Fig. 2.6. Bases for Game of life: $U$, $V$ for the von Neumann type, $W$ for the Moore type

In the case shown in Figs. 6.1, 6.2 the cellular c-e structures $U$, $V$, $W$ are embedded into identical Cartesian coordinate system. In each of them, the coordinates of any cell differ from coordinates of its neighbours, at most by one. Important is that this difference is <u>the same</u> in both components of summation. Otherwise, the summation yields incorrect result, i.e. not a <u>cellular</u> c-e structure. This is shown in Fig. 6.3, where the <u>cellular</u> c-e structure $U$ has been modified to $U'$, so that its cells have both even coordinates, the <u>cellular</u> c-e structure $V$ is as in Fig. 6.1, but $W$ becomes $W'$ thus <u>not cellular</u> c-e structure, because its cells with both even coordinates have eight neighbours, whereas remaining only four. Notice that in accordance with the concept of substructure (Chap. 1) and partial order between c-e structures, the cellular c-e structures $U$, $V$ are substructures of $W$, i.e. $U \leq W$ and $V \leq W$.

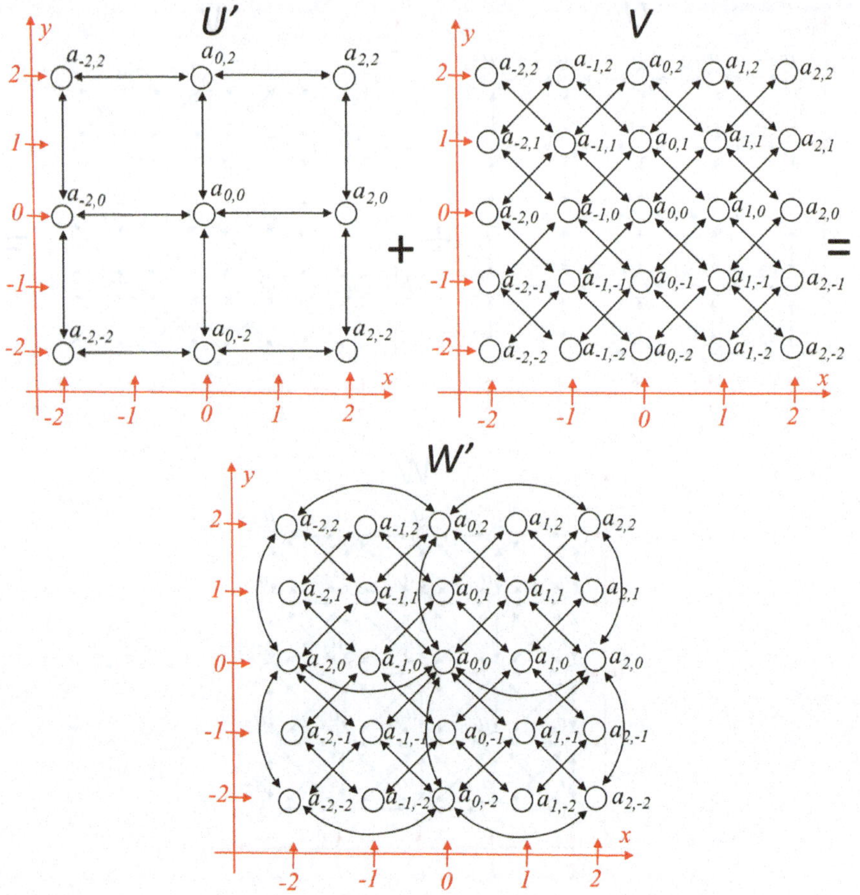


**Fig. 6.3** Fragments of cellular c-e structures $U'$, $V$ and the outcome of summation $W' = U' + V$, being not a cellular c-e structure

Also, the summation of $U'$ from Fig. 6.3 with $U$ from Fig. 6.1 yields an ordinary c-e structure $W''$, not a cellular one, as shows Fig. 6.4.

## Remark

The arrangement of nodes in the c-e structures $W$, in Figs. 6.1, 6.2 on Cartesian plane is the same as of octagonal tiles-cells in Fig. 2.6 (Chap. 2) that do not tessellate the plane. Similarly, the plane cannot be tessellated by octagonal and square tiles hosting nodes of c-e structures ($W'$, $W''$ not cellular) with retained interconnection of cells such as in Fig. 2.6. Surely, treating all cells as octagonal and square tiles sharing sides, one can cover the plane surface by them, without gaps and overlaps, i.e. to

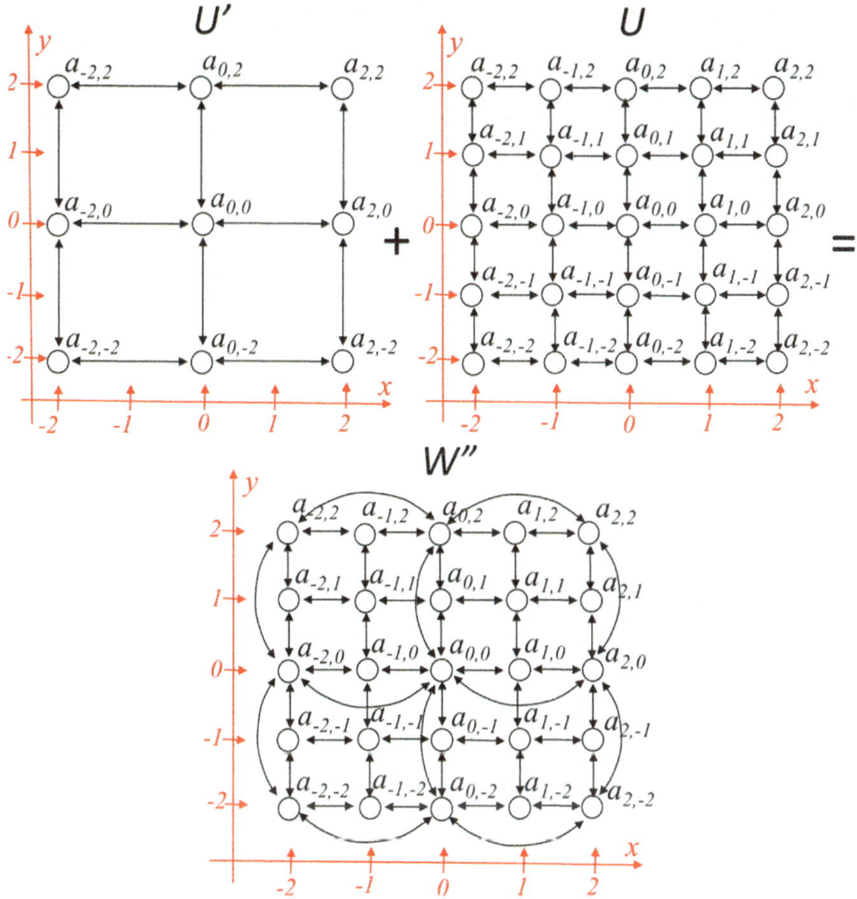


**Fig. 6.4** Fragments of cellular c-e structures $U'$, $U$ and the outcome of summation $W'' = U' + U$, being not a cellular c-e structure

tessellate the plane as shows Fig. 6.5. But some interconnections of cells in Fig. 6.1 (so, in Fig. 2.6) would be broken: such c-e structure, becomes not cellular. Note also, that sometimes attempts of tiling the plane may cause disappearance of some nodes, as shows Fig. 2.7 (Chap. 2).

Now, let us consider how to get a desirable semantics of the sum of cellular c-e structures, given semantics of the components. To this end, encoding of next-state rules as tables of functions making parameters $B_{\mathbf{j}\to\mathbf{i}}$ will be applied. Recall that columns of the table $B_{\mathbf{j}\to\mathbf{i}}$ correspond to neighbours surrounding a cell $a_{x,y}$ and rows correspond to functions $f_k^{\mathbf{j}\to\mathbf{i}} : nbh(a_{x,y}) \overset{\text{into}}{\to} \mathbb{Q}$ that make change value of $a_{x,y}$ from **j** to **i**. Remember also that neighbours surround each cell in the same order (all atoms are isomorphic). As an example of combining the next-state rules, let us take two

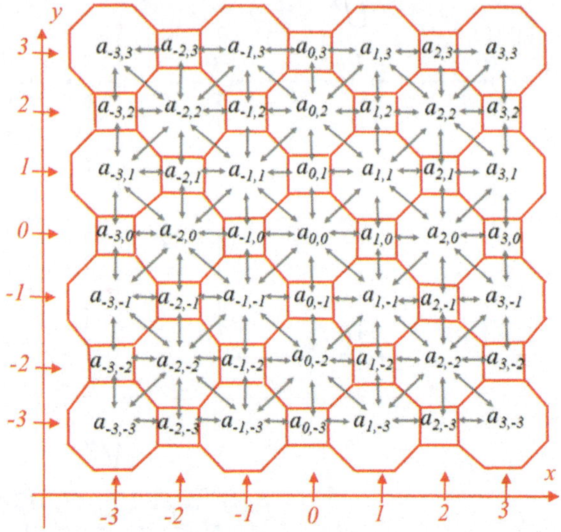


**Fig. 6.5** Tiling the plane by octagons and squares hosting cells. The square tiles have 4 neighbours, the octagonal have them 8: not a cellular c-e structure. Some interconnections between the cells have been lost as compared with Fig. 6.1

**Table 6.1** In the rows are values of functions $f_k^{0\to1} \in B_{0\to1}^{birth\text{-}at\text{-}U}$ of domain $\underset{U}{nbh}(a_{x,y})$ for the cellular c-e structure $U$ in Fig. 6.1 defining birth in the cell $a_{x,y}$ at $U$

| $\underset{U}{nbh}(a_{x,y})$ : | $a_{x-1,y}$ | $a_{x,y+1}$ | $a_{x+1,y}$ | $a_{x,y-1}$ |
|---|---|---|---|---|
| $f_1^{0\to1}$ | **1** | **1** | **1** | **0** |
| $f_2^{0\to1}$ | **1** | **1** | **0** | **1** |
| $f_3^{0\to1}$ | **1** | **0** | **1** | **1** |
| $f_4^{0\to1}$ | **0** | **1** | **1** | **1** |

geometric arrangements $U$ and $V$ of the Game of life of the von Neumann type, depicted in Fig. 6.1.

The parameters $B_{0\to1}^{birth-at-U} = \{f_1^{0\to1}, f_2^{0\to1}, f_3^{0\to1}, f_4^{0\to1}\}$ and $B_{1\to1}^{surv\text{-}at\text{-}U} = \{f_1^{1\to1}, f_2^{1\to1}, f_3^{1\to1}, f_4^{1\to1}, f_5^{1\to1}, f_6^{1\to1}, f_7^{1\to1}, f_8^{1\to1}, f_9^{1\to1}, f_{10}^{1\to1}\}$ that define semantic rule for $U$ are given by means of Tables 6.1, 6.2 and for $V$ by means of Tables 6.3, 6.4. The remaining parameters $B_{\mathrm{j}\to\mathrm{i}}$ are evident—see Chap. 5 (Fig. 5.1).

The tables "birth" and "survival" defining the semantic rule for $W$, the cellular c-e structure of the Game of life of the Moore type, are much larger: $\binom{8}{3} = 56$ rows for "birth" and $\binom{8}{2} + \binom{8}{3} = 28 + 56 = 84$ rows for "survival". Table 6.5 contains first 13 out of 56 patterns $f_k''^{0\to1}$ $(k = 1, \ldots, 56)$ defining conditions for "birth" at the cellular structure $W$. Table 6.6 contains 12 out of 84 patterns $f_k''^{1\to1}$ $(k = 1, \ldots, 84)$ defining conditions for "survival" at the cellular structure $W$.

**Table 6.2** In the rows are values of functions $f_k^{1\to 1} \in B_{1\to 1}^{surv-at-U}$ of domain $\underset{U}{nbh}(a_{x,y})$ for the cellular c-e structure $U$ in Fig. 6.1 defining survival in the cell $a_{x,y}$ at $U$

| $\underset{U}{nbh}(a_{x,y})$ : | $a_{x-1,y}$ | $a_{x,y+1}$ | $a_{x+1,y}$ | $a_{x,y-1}$ |
|---|---|---|---|---|
| $f_1^{1\to 1}$ | 0 | 0 | 1 | 1 |
| $f_2^{1\to 1}$ | 0 | 1 | 1 | 0 |
| $f_3^{1\to 1}$ | 1 | 1 | 0 | 0 |
| $f_4^{1\to 1}$ | 0 | 1 | 0 | 1 |
| $f_5^{1\to 1}$ | 1 | 0 | 1 | 0 |
| $f_6^{1\to 1}$ | 1 | 0 | 0 | 1 |
| $f_7^{1\to 1}$ | 1 | 1 | 1 | 0 |
| $f_8^{1\to 1}$ | 1 | 1 | 0 | 1 |
| $f_9^{1\to 1}$ | 1 | 0 | 1 | 1 |
| $f_{10}^{1\to 1}$ | 0 | 1 | 1 | 1 |

**Table 6.3** In the rows are values of functions $f_k'^{0\to 1} \in B_{0\to 1}^{birth-at-V}$ of domain $\underset{V}{nbh}(a_{x,y})$ for the cellular c-e structure $V$ in Fig. 6.1 defining birth in the cell $a_{x,y}$ at $V$

| $\underset{V}{nbh}(a_{x,y})$ : | $a_{x-1,y-1}$ | $a_{x-1,y+1}$ | $a_{x+1,y+1}$ | $a_{x+1,y-1}$ |
|---|---|---|---|---|
| $f_1'^{0\to 1}$ | 1 | 1 | 1 | 0 |
| $f_2'^{0\to 1}$ | 1 | 1 | 0 | 1 |
| $f_3'^{0\to 1}$ | 1 | 0 | 1 | 1 |
| $f_4'^{0\to 1}$ | 0 | 1 | 1 | 1 |

**Table 6.4** In the rows are values of functions $f_k'^{1\to 1} \in B_{1\to 1}^{surv-at-V}$ of domain $\underset{V}{nbh}(a_{x,y})$ for the cellular c-e structure $V$ in Fig. 6.1 defining survival in the cell $a_{x,y}$ at $V$

| $\underset{V}{nbh}(a_{x,y})$ : | $a_{x-1,y-1}$ | $a_{x-1,y+1}$ | $a_{x+1,y+1}$ | $a_{x+1,y-1}$ |
|---|---|---|---|---|
| $f_1'^{1\to 1}$ | 0 | 0 | 1 | 1 |
| $f_2'^{1\to 1}$ | 0 | 1 | 1 | 0 |
| $f_3'^{1\to 1}$ | 1 | 1 | 0 | 0 |
| $f_4'^{1\to 1}$ | 0 | 1 | 0 | 1 |
| $f_5'^{1\to 1}$ | 1 | 0 | 1 | 0 |
| $f_6'^{1\to 1}$ | 1 | 0 | 0 | 1 |
| $f_7'^{1\to 1}$ | 1 | 1 | 1 | 0 |
| $f_8'^{1\to 1}$ | 1 | 1 | 0 | 1 |
| $f_9'^{1\to 1}$ | 1 | 0 | 1 | 1 |
| $f_{10}'^{1\to 1}$ | 0 | 1 | 1 | 1 |

**Table 6.5** In the 56 rows are values of functions $f_k''^{\,0\to1} \in B_{0\to1}^{birth-at-W}$ of domain ${}^{nbh}_{W}(a_{x,y})$ for the cellular c-e structure $W = U + V$ in Fig. 6.1 defining "birth" in the cell $a_{x,y}$ at $W$

| $nbh(a_{x,y})$: $W$ | $a_{x-1,y-1}$ | $a_{x-1,y}$ | $a_{x-1,y+1}$ | $a_{x,y+1}$ | $a_{x+1,y+1}$ | $a_{x+1,y}$ | $a_{x+1,y-1}$ | $a_{x,y-1}$ |
|---|---|---|---|---|---|---|---|---|
| $f''_1{}^{0\to1}$ | 0 | 1 | 0 | 1 | 0 | 1 | 0 | 0 |
| $f''_2{}^{0\to1}$ | 0 | 1 | 0 | 1 | 0 | 0 | 0 | 1 |
| $f''_3{}^{0\to1}$ | 0 | 1 | 0 | 0 | 0 | 1 | 0 | 1 |
| $f''_4{}^{0\to1}$ | 0 | 0 | 0 | 1 | 0 | 1 | 0 | 1 |
| $f''_5{}^{0\to1}$ | 1 | 0 | 1 | 0 | 1 | 0 | 0 | 0 |
| $f''_6{}^{0\to1}$ | 1 | 0 | 1 | 0 | 0 | 0 | 1 | 0 |
| $f''_7{}^{0\to1}$ | 1 | 0 | 0 | 0 | 1 | 0 | 1 | 0 |
| $f''_8{}^{0\to1}$ | 0 | 0 | 1 | 0 | 1 | 0 | 1 | 0 |
| $f''_9{}^{0\to1}$ | 0 | 1 | 0 | 1 | 0 | 0 | 0 | 1 |
| $f''_{10}{}^{0\to1}$ | 1 | 0 | 0 | 0 | 1 | 0 | 1 | 0 |
| $f''_{11}{}^{0\to1}$ | 1 | 0 | 1 | 0 | 0 | 0 | 1 | 0 |
| $f''_{12}{}^{0\to1}$ | 0 | 1 | 0 | 0 | 0 | 1 | 0 | 1 |
| $f''_{13}{}^{0\to1}$ | 0 | 0 | 1 | 0 | 1 | 0 | 1 | 0 |

**Table 6.6** In the 84 rows are values of functions $f_k''^{\,1\to1} \in B_{1\to1}^{surv-at-W}$ of domain ${}^{nbh}_{W}(a_{x,y})$ for the cellular c-e structure $W$ in Fig. 6.1 defining "survival" in the cell $a_{x,y}$ at $W$.

| $nbh(a_{x,y})$: $W$ | $a_{x-1,y-1}$ | $a_{x-1,y}$ | $a_{x-1,y+1}$ | $a_{x,y+1}$ | $a_{x+1,y+1}$ | $a_{x+1,y}$ | $a_{x+1,y-1}$ | $a_{x,y-1}$ |
|---|---|---|---|---|---|---|---|---|
| $f''_1{}^{1\to1}$ | 0 | 1 | 0 | 1 | 0 | 1 | 0 | 0 |
| $f''_2{}^{1\to1}$ | 0 | 1 | 0 | 1 | 0 | 0 | 0 | 1 |
| $f''_3{}^{1\to1}$ | 0 | 1 | 0 | 0 | 0 | 1 | 0 | 1 |
| $f''_4{}^{1\to1}$ | 0 | 0 | 0 | 1 | 0 | 1 | 0 | 1 |
| $f''_5{}^{1\to1}$ | 1 | 0 | 1 | 0 | 1 | 0 | 0 | 0 |
| $f''_6{}^{1\to1}$ | 1 | 0 | 1 | 0 | 0 | 0 | 1 | 0 |
| $f''_{35}{}^{1\to1}$ | 0 | 0 | 0 | 0 | 0 | 1 | 0 | 1 |
| $f''_{36}{}^{1\to1}$ | 0 | 0 | 0 | 1 | 0 | 1 | 0 | 0 |
| $f''_{37}{}^{1\to1}$ | 0 | 1 | 0 | 1 | 0 | 0 | 0 | 0 |
| $f''_{38}{}^{1\to1}$ | 0 | 0 | 0 | 0 | 1 | 0 | 1 | 0 |
| $f''_{39}{}^{1\to1}$ | 0 | 0 | 1 | 0 | 1 | 0 | 0 | 0 |
| $f''_{40}{}^{1\to1}$ | 1 | 0 | 0 | 0 | 0 | 0 | 0 | 1 |

In order to exemplify usage of such semantic tables, let for instance, the cellular c-e structure implementing the Moore type Game of life be in a state $s$ such that $s(a_{x,y}) = \mathbf{0}$ ("dead" cell $a_{x,y}$) and let the next state be $t$, that is let $(s, t) \in [\![a_{x,y}]\!]$ hold. Looking for a status (value) of $a_{x,y}$ in the state $t$ one has to check if $s|_{nbh(a_{x,y})} \in B^{birth-at-W}_{\mathbf{0}\rightarrow\mathbf{1}}$, that is if the Table 6.5 contains function (pattern) $f''^{\mathbf{0}\rightarrow\mathbf{1}}_k$ identical with restriction of $s$ to $\underset{W}{nbh}(a_{x,y})$. If yes then $t(a_{x,y}) = \mathbf{1}$, because according to definition of semantics, the formula

$$s(a_{x,y}) = \mathbf{0} \wedge s|_{nbh(a_{x,y})} \in B^{birth-at-W}_{\mathbf{0}\rightarrow\mathbf{1}} \wedge t(a_{x,y}) = \mathbf{1}$$

holds, otherwise, i.e. if $s|_{nbh(a_{x,y})} \notin B^{birth-at-W}_{\mathbf{0}\rightarrow\mathbf{1}}$ then $t(a_{x,y}) = \mathbf{0}$. Similarly one may check whether or not the "live" cell in the state $s$ ($s(a_{x,y}) = \mathbf{1}$) would "survive" in the next state $t$, by testing the Table 6.6 i.e. $B^{surv-at-W}_{\mathbf{1}\rightarrow\mathbf{1}}$. See the multigraph in Fig. 5.1 Chap. 5).

The above examples reveal some problems with combining cellular c-e structures by their summation, both at the structural and semantic side. At the structural side, the well defined conditions should guarantee equinumerous neighbourhood of all cells in the result. Otherwise some interconnections of cells disappear in the result, what happened, e.g. in Figs. 6.3, 6.4. At the semantic side, a requirement is that in constructing table for the result of summation, out of respective tables for components, no conflicts of values on entries occur, and that the constructed tables $B_{\mathbf{j}\rightarrow\mathbf{i}}$ (containing functions $f^{\mathbf{j}\rightarrow\mathbf{i}}_k$) determine intended behaviour of the result. Consider, for instance, building the Moore type Game of life $W$ out of two versions $U$ and $V$ of the von Neumann type (Fig. 6.1). In constructing Table 6.5 out of Tables 6.1, 6.3, no conflicts were possible. Similarly, in constructing Table 6.6 out of Tables 6.2, 6.4. This was due to disjoint domains of functions $f^{\mathbf{0}\rightarrow\mathbf{1}}_k$ in Tables 6.1, 6.3, as well as of functions $f^{\mathbf{1}\rightarrow\mathbf{1}}_k$ in Tables 6.2, 6.4. In the Game of life, only a number of active ("live") neighbours of a cell matters, not their position (coordinates). A semantics of the result $W$ of addition $U + V$ is accomplished by tables whose certain truncations yield tables for $U$ and $V$. Precisely, the Table 6.5 (birth at $W$) is such one, that removing from it:

- columns $a_{x-1,y-1}, a_{x-1,y+1}, a_{x+1,y+1}, a_{x+1,y-1}$ (i.e. $\underset{V}{nbh}(a_{x,y})$), and then rows that correspond to functions $f''^{\mathbf{0}\rightarrow\mathbf{1}}_k$ whose restrictions to $\underset{U}{nbh}(a_{x,y})$ are absent in the Table 6.1, birth at $U$ is obtained,
- columns $a_{x-1,y}, a_{x,y+1}, a_{x+1,y}, a_{x,y-1}$, (i.e. $\underset{U}{nbh}(a_{x,y})$), and then rows that correspond to functions $f''^{\mathbf{0}\rightarrow\mathbf{1}}_k$ whose restrictions to $\underset{V}{nbh}(a_{x,y})$ are absent in the Table 6.3, birth at $V$ is obtained.

Analogously the Table 6.6 that accomplishes survival at $W$ can be specified.

Construction of semantic tables "birth" and "survival" of the von Neumann and Moore type, has been done without specifying initial state of the game. A concrete course of the game is determined by these tables when an initial state is provided.

Before some other problems of combining semantics are considered, let us take a look at combination of semantics of cellular c-e structures generating red, blue and green light rays shown in Fig. 5.2 (Chap. 5). Here, a white (value **0**) cell $a_{x,y}$ turns to red (value **1**) iff its neighbour $a_{x-1,y+1}$ or $a_{x+1,y-1}$ is yellow (value $\omega$) or red and remaining neighbours are white; a white cell $a_{x,y}$ turns to blue (value **2**) iff its neighbour $a_{x-1,y-1}$ or $a_{x+1,y+1}$ is yellow or blue and remaining neighbours are white; a white cell $a_{x,y}$ turns to green (value **3**) iff its neighbour $a_{x,y+2}$ or $a_{x,y-2}$ is yellow or green and remaining neighbours are white. The red, blue, green cells do not change colour. Assuming that the cellular c-e structure $U_{red} + U_{blue} + U_{green}$ starts evolution from the initial state $s_{init}$ with $s_{init}(a_{0,0}) = \omega$ and $s_{init}(a_{x,y}) = \mathbf{0}$ for remaining cells $a_{x,y}$, an outcome of its (partial) evolution shows Fig. 6.6. In this task, unlike the Game of life, the distribution of neighbours holding specific values for intended behaviour of a cell, is the same for each cell. The components $U_{red}$, $U_{blue}$, $U_{green}$ are structurally identical but with different next-state rules incorporating Tables 5.3, 5.4, 5.5 (Chap. 5). The cellular c-e structure $U_{rbg} = U_{red} + U_{blue} + U_{green}$ works under their combination shown in Table 6.7 with parameters $B^{rbg}_{\mathbf{0}\to\mathbf{i}}$, obtained by "downward concatenation" of these tables, that is the table containing all rows of the latter.

Remaining parameters $B^{rbg}_{\mathbf{j}\to\mathbf{i}}$ involved in creation of the sum $U_{rbg} = U_{red} + U_{blue} + U_{green}$ are: $B^{rbg}_{\mathbf{1}\to\mathbf{1}} = B^{rbg}_{\mathbf{2}\to\mathbf{2}} = B^{rbg}_{\mathbf{3}\to\mathbf{3}} = B^{rbg}_{\omega\to\omega} = \{\mathbf{1}, \mathbf{2}, \mathbf{3}, \omega\}^{nbh(a_{x,y})}$, thus if $a_{x,y}$ is red, blue, green, yellow, then it never changes colour (value), because $s|_{nbh(a_{x,y})} \in B^{rbg}_{\mathbf{i}\to\mathbf{i}}$ ($\mathbf{i} = \mathbf{1}, \mathbf{2}, \mathbf{3}, \omega$) is always true. Besides $B^{rbg}_{\mathbf{0}\to\mathbf{0}} = \overline{B^{rbg}_{\mathbf{0}\to\mathbf{i}}}$ ($\mathbf{i} = \mathbf{1}, \mathbf{2}, \mathbf{3}, \omega$) thus if $a_{x,y}$ is white then it remains white when $s|_{nbh(a_{x,y})} \notin B^{rbg}_{\mathbf{0}\to\mathbf{i}}$ (i.e. function $s|_{nbh(a_{x,y})}$ does not

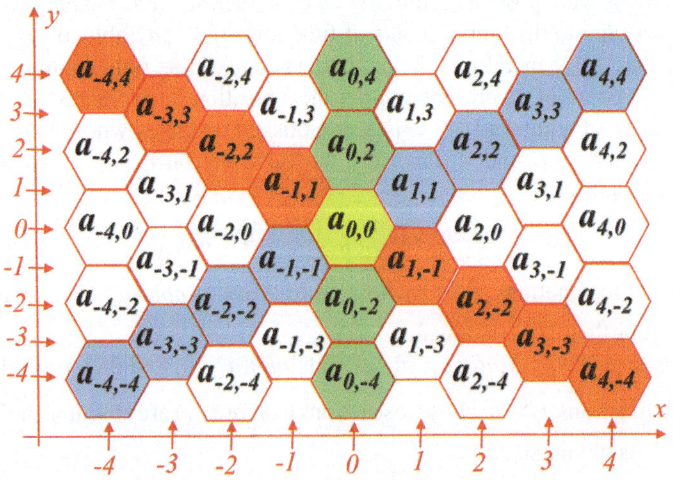


**Fig. 6.6** Image of light rays propagating from the yellow cell $a_{0,0}$ and generated by the cellular c-e structure placed on the *board*6 (Fig. 2.9 in Chap. 2) if initial state $s$ is $s(a_{0,0}) = \omega$ and $s(a_{x,y}) = \mathbf{0}$ for remaining $x$, $y$. This is the sum $U_{rbg} = U_{red} + U_{blue} + U_{green}$ of cellular c-e structures shown in Fig. 5.2 (Chap. 5)

**Table 6.7** Integrated table $B^{rbg}_{0\rightarrow i}$ out of three Tables 5.3, 5.4, 5.5 (Chap. 5) defining sets $B^{red}_{0\rightarrow 1} = \{f_1^{0\rightarrow 1}, f_2^{0\rightarrow 1}, f_3^{0\rightarrow 1}, f_4^{0\rightarrow 1}\}$, $B^{blue}_{0\rightarrow 2} = \{f_1^{0\rightarrow 2}, f_2^{0\rightarrow 2}, f_3^{0\rightarrow 2}, f_4^{0\rightarrow 2}\}$, $B^{green}_{0\rightarrow 3} = \{f_1^{0\rightarrow 3}, f_2^{0\rightarrow 3}, f_3^{0\rightarrow 3}, f_4^{0\rightarrow 3}\}$ of functions with the same domain $nbh(a_{x,y})$ in the top line and with values in the rows underneath. The table defines formation of six light rays simultaneously, depicted in Fig. 6.6 if $U_{rbg}$ starts from initial state $s$ such that $s(a_{0,0}) = \omega$ and $s(a_{x,y}) = \mathbf{0}$ for remaining $x, y$

| | $nbh(a_{x,y})$: | $a_{x-1,y-1}$ | $a_{x-1,y+1}$ | $a_{x,y+2}$ | $a_{x+1,y+1}$ | $a_{x+1,y-1}$ | $a_{x,y-2}$ |
|---|---|---|---|---|---|---|---|
| $B^{red}_{0\rightarrow 1}$ | $f_1^{0\rightarrow 1}$ | 0 | 1 | 0 | 0 | 0 | 0 |
| | $f_2^{0\rightarrow 1}$ | 0 | 0 | 0 | 0 | 1 | 0 |
| | $f_3^{0\rightarrow 1}$ | 0 | ω | 0 | 0 | 0 | 0 |
| | $f_4^{0\rightarrow 1}$ | 0 | 0 | 0 | 0 | ω | 0 |
| $B^{blue}_{0\rightarrow 2}$ | $f_1^{0\rightarrow 2}$ | 2 | 0 | 0 | 0 | 0 | 0 |
| | $f_2^{0\rightarrow 2}$ | 0 | 0 | 0 | 2 | 0 | 0 |
| | $f_3^{0\rightarrow 2}$ | ω | 0 | 0 | 0 | 0 | 0 |
| | $f_4^{0\rightarrow 2}$ | 0 | 0 | 0 | ω | 0 | 0 |
| $B^{green}_{0\rightarrow 3}$ | $f_1^{0\rightarrow 3}$ | 0 | 0 | 3 | 0 | 0 | 0 |
| | $f_2^{0\rightarrow 3}$ | 0 | 0 | 0 | 0 | 0 | 3 |
| | $f_3^{0\rightarrow 3}$ | 0 | 0 | ω | 0 | 0 | 0 |
| | $f_4^{0\rightarrow 3}$ | 0 | 0 | 0 | 0 | 0 | ω |

appear in the Table 6.7) and $B^{rbg}_{0\rightarrow\omega} = B^{rbg}_{1\rightarrow 0} = B^{rbg}_{1\rightarrow 2} = B^{rbg}_{1\rightarrow 3} = B^{rbg}_{1\rightarrow\omega} = B^{rbg}_{2\rightarrow 0} = B^{rbg}_{2\rightarrow 1} = B^{rbg}_{2\rightarrow 3} = B^{rbg}_{2\rightarrow\omega} = B^{rbg}_{3\rightarrow 0} = B^{rbg}_{3\rightarrow 1} = B^{rbg}_{3\rightarrow 2} = B^{rbg}_{3\rightarrow\omega} = B^{rbg}_{\omega\rightarrow 0} = B^{rbg}_{\omega\rightarrow 1} = B^{rbg}_{\omega\rightarrow 2} = B^{rbg}_{\omega\rightarrow 3} = \varnothing$.

The Tables 5.3, 5.4, 5.5 specified separately propagation of each ray. The Table 6.7 specifies propagation of all rays simultaneously, due to the parallel semantics of cellular c-e structures. In accordance with general semantic rule in Chap. 5, the relation $[\![a_{x,y}]\!]$ that determines behaviour of a cell $a_{x,y}$ is defined as:

$(s, t) \in [\![a_{x,y}]\!]$ holds if and only if certain of the following formulas holds:

$$\left[\bigvee_{\mathbf{j}\in\{\mathbf{0},\mathbf{1},\mathbf{2},\mathbf{3},\omega\}} s(a_{x,y}) = \mathbf{j} \wedge s|_{nbh(a_{x,y})} \subset B^{rbg}_{\mathbf{j}\rightarrow\mathbf{0}}\right] \wedge t(a_{x,y}) = \mathbf{0}$$

$$\left[\bigvee_{\mathbf{j}\in\{\mathbf{0},\mathbf{1},\mathbf{2},\mathbf{3},\omega\}} s(a_{x,y}) = \mathbf{j} \wedge s|_{nbh(a_{x,y})} \in B^{rbg}_{\mathbf{j}\rightarrow\mathbf{1}}\right] \wedge t(a_{x,y}) = \mathbf{1}$$

$$\left[\bigvee_{\mathbf{j}\in\{\mathbf{0},\mathbf{1},\mathbf{2},\mathbf{3},\omega\}} s(a_{x,y}) = \mathbf{j} \wedge s|_{nbh(a_{x,y})} \in B^{rbg}_{\mathbf{j}\rightarrow\mathbf{2}}\right] \wedge t(a_{x,y}) = \mathbf{2}$$

$$\left[\bigvee_{\mathbf{j}\in\{\mathbf{0},\mathbf{1},\mathbf{2},\mathbf{3},\omega\}} s(a_{x,y}) = \mathbf{j} \wedge s|_{nbh(a_{x,y})} \in B^{rbg}_{\mathbf{j}\rightarrow\mathbf{3}}\right] \wedge t(a_{x,y}) = \mathbf{3}$$

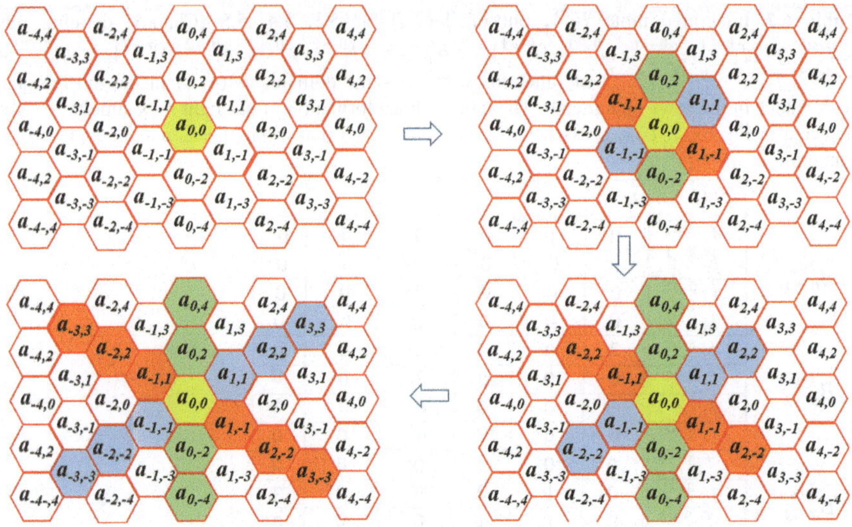


**Fig. 6.7** Evolution of cellular c-e structure $U_{rbg} = U_{red} + U_{blue} + U_{green}$ – propagation of six light rays

$$\left[\bigvee_{\mathbf{j}\in\{\mathbf{0},\mathbf{1},\mathbf{2},\mathbf{3},\omega\}} s(a_{x,y}) = \mathbf{j} \wedge s|_{nbh(a_{x,y})} \in B^{rbg}_{\mathbf{j}\to\omega}\right] \wedge t(a_{x,y}) = \omega$$

In words: $(s, t) \in [\![a_{x,y}]\!]$ holds when $s(a_{x,y}) = \mathbf{j}, t(a_{x,y}) = \mathbf{i}\,(\mathbf{j}, \mathbf{i} = \mathbf{0}, \mathbf{1}, \mathbf{2}, \mathbf{3}, \omega)$ and a restriction of the state $s$ to $nbh(a_{x,y})$ exists in a set $B^{rbg}_{\mathbf{0}\to\mathbf{i}} = B^{red}_{\mathbf{0}\to\mathbf{i}} \cup B^{blue}_{\mathbf{0}\to\mathbf{i}} \cup B^{green}_{\mathbf{0}\to\mathbf{i}} \cup B^{yellow}_{\mathbf{0}\to\mathbf{i}}$ of functions $f_k^{\mathbf{0}\to\mathbf{i}}$ ($\mathbf{i} = 1, 2, 3$, $k = 1, 2, 3, 4$) collected in Table 6.7, whereas remaining parameters (sets $B^{rbg}_{\mathbf{j}\to\mathbf{i}}$) are defined above.

Four successive stages of evolution of cellular c-e structure $U_{rbg} = U_{red} + U_{blue} + U_{green}$ starting with the single cell $a_{0,0}$ with $s(a_{0,0}) = \omega$ and semantics specified by above parameters $B^{rbg}_{\mathbf{j}\to\mathbf{i}}$ ($\mathbf{j}$,$\mathbf{i} = \mathbf{0}, \mathbf{1}, \mathbf{2}, \mathbf{3}, \omega$) is shown in Fig. 6.7

### Remark

Combining c-e structures into cellular ones has been essentially exploited in Chap. 3, where flat bases (grounds) hosting 2-dimensional structures were transformed ("bent") into surfaces of solid objects like cylinders and toruses, in general—manifolds. The components of such combining were non-cellular c-e structures. Examples in Figs. 6.3, 6.4 show that also conversely: combining cellular c-e structures may result in non-cellular. So, various cases one may encounter. The next chapter contains investigation of conditions for structural and semantical correctness in effect of combining.

# Chapter 7
# Correct and Incorrect Summation

The von Neumann type Game of life cellular c-e structures $U$, $V$ and of the Moore type $W$ in Fig. 6.2 (Chap. 6), have identical set of cells, but the same cell in each of them has different ($U$ and $V$—disjoint) neighbourhood. Thus, $U$, $V$, $W$ are structurally different. Nonetheless, behaviourally identical, in accordance with the global semantic rule for this task, because any cell $a_{x,y}$, shared by $U$, $V$, $W$, modifies its value (dead/live, i.e. **0/1**) the same way, depending only on the number of live neighbours, not of the location of live neighbours in the neighbourhood of $a_{x,y}$. This is not so with cellular c-e structures generating separately the red, blue, green rays—Fig. 5.2 (Chap. 5), then combined into one cellular c-e structure (Fig. 6.6), which generates all these rays simultaneously. They have identical set of cells too and any cell shared in each of them has the same neighbourhood: they all are structurally identical. But these cellular c-e structures are behaviourally different, because the same cell $a_{x,y}$ in $U$ and $V$ may modify its value diversely, depending on current distribution of values $(\mathbf{0}, \mathbf{1}, \mathbf{2}, \mathbf{3}, \omega)$[1] of this cell's neighbours in individual structure. Here, the semantic rule is different for each of these cellular c-e structures. Each performs different task—emits light rays from $a_{0,0}$ in a direction specified by its rule. The multigraph in Fig. 7.1 shows value changes of $a_{x,y}$ in the combined cellular c-e structure in Fig. 6.6. The change from **0** to **1** or to **2** or to **3** in a state $s$ occurs iff restricted function $s|_{nbh(a_{x,y})}$ is in the Table 6.7, otherwise $a_{x,y}$ retains value **0** in the next state $t$.

### Remark

Obviously, for the cellular c-e structures in Fig. 5.2 (Chap. 5) there is no need to provide explicitly tables of patterns for parameters $B^{red}_{\mathbf{1}\to\mathbf{1}}$, $B^{blue}_{\mathbf{2}\to\mathbf{2}}$, $B^{green}_{\mathbf{3}\to\mathbf{3}}$, $B^{yellow}_{\omega\to\omega}$ because if $s(a_{x,y})$ is **1**, **2**, **3**, $\omega$ then the formulas $s|_{nbh(a_{x,y})} \in B^{red}_{\mathbf{1}\to\mathbf{1}}$, $s|_{nbh(a_{x,y})} \in B^{blue}_{\mathbf{2}\to\mathbf{2}}$, $s|_{nbh(a_{x,y})} \in B^{green}_{\mathbf{3}\to\mathbf{3}}$, $s|_{nbh(a_{x,y})} \in B^{yellow}_{\omega\to\omega}$ in definition of the next-state rule for $a_{x,y}$, are fulfilled. Note that there are no transits $\mathbf{1}\to\mathbf{0}$, $\mathbf{1}\to\mathbf{2}$, $\mathbf{1}\to\mathbf{3}$, $\mathbf{1}\to\omega$, $\mathbf{2}\to\mathbf{0}$, $\mathbf{2}\to\mathbf{1}$, $\mathbf{2}\to\mathbf{3}$, $\mathbf{2}\to\omega$, $\mathbf{3}\to\mathbf{0}$, $\mathbf{3}\to\mathbf{1}$, $\mathbf{3}\to\mathbf{2}$, $\mathbf{3}\to\omega$, $\omega\to\mathbf{0}$, $\omega\to\mathbf{1}$, $\omega\to\mathbf{2}$,

---

[1] **0**-white, **1**-red, **2**-blue, **3**-green, $\omega$-yellow.

 
L. Czaja, *Cellular Cause-Effect Structures*, Studies in Systems, Decision and Control 526,
https://doi.org/10.1007/978-3-031-55590-9_7

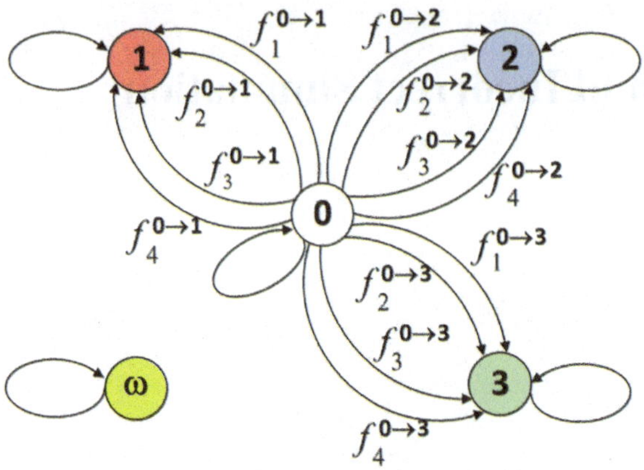


**Fig. 7.1** Multigraph-automaton of behaviour of the cell $a_{x,y}$ in the cellular c-e structure producing light rays, shown in Fig. 6.6. The edges $\mathbf{0} \to \mathbf{1}$, $\mathbf{0} \to \mathbf{2}$, $\mathbf{0} \to \mathbf{3}$ are labelled with functions respectively: $\{f_1^{0\to1}, f_2^{0\to1}, f_3^{0\to1}, f_4^{0\to1}\}$, $\{f_1^{0\to2}, f_2^{0\to2}, f_3^{0\to2}, f_4^{0\to2}\}$, $\{f_1^{0\to3}, f_2^{0\to3}, f_3^{0\to3}, f_4^{0\to3}\}$, collected in the table $B^{rbg}_{\mathbf{0}\to\mathbf{i}}$ (Table 6.7, Chap. 6). The self-loops $\mathbf{1} \to \mathbf{1}$, $\mathbf{2} \to \mathbf{2}$, $\mathbf{3} \to \mathbf{3}$, $\omega \to \omega$ never change state of the automaton (status of $a_{x,y}$) as if labelled with any function of type $\{\mathbf{0}, \mathbf{1}, \mathbf{2}, \mathbf{3}, \omega\}^{nbh(a_{x,y})}$. The self-loop $\mathbf{0} \to \mathbf{0}$ retains status $\mathbf{0}$ of $a_{x,y}$ in a state $s$, only if no function from the Table 6.7 matches the restricted state $s|_{nbh(a_{x,y})}$, that is the state of the neighbourhood of $a_{x,y}$

$\omega \to \mathbf{3}$, because respective sets $B^{red}_{\mathbf{1}\to\mathbf{0}}$, $B^{red}_{\mathbf{1}\to\mathbf{2}}$, $B^{red}_{\mathbf{1}\to\mathbf{3}}$, $B^{red}_{\mathbf{1}\to\omega}$, $B^{blue}_{\mathbf{2}\to\mathbf{0}}$, $B^{blue}_{\mathbf{2}\to\mathbf{1}}$, $B^{blue}_{\mathbf{2}\to\mathbf{3}}$, etc. are empty, see the multigraph in Fig. 7.1.

As shown above, the summation of cellular c-e structures not always yields correct result—in contrast to summing (and multiplying) general c-e structures. Before coming to conditions ensuring such correctness or its lack, let us take a look at the notion of *neighbourhood*—a set required to be equinumerous for each cell in the result of summation. If cells are perceived as geometric shapes (manifolds in Cartesian space), then their neighbouring, meant as adherence of sides of polygons or faces of polyhedrons, does not conform to the meaning of "neighbour" adopted in cellular cause-effect structures.[2] Nor the "neighbouring" refers here to the notion like "the least distance" between neighbouring cells, absent notion in cellular automata[3] and in cellular c-e structures. The cells are vertices in a graph and are neighbours when linked by a bidirectional edge, i.e. when one cell is the cause and effect of the other.[4] Remember that in the considered geometric model, all structures reside at the same Cartesian coordinate system.

---

[2] Nor in the cellular automata.

[3] Although the notion of a "neighbourhood radius" has been introduced in cellular automata, but not in their cause-effect version.

[4] The diversity of the meaning of "neighbour" manifests itself also in the chess game, where the neighbouring square for the knight and for the king are different places for their permitted move. Note that the knight and king may be seen as current status of a place at the chessboard, i.e. values of state.

## 7.1 Structural Correctness

The requirement for structurally correct summation of cellular c-e structures is that all atoms in the result be equivalent relative-to shift transformation, which implies equinumerous neighbourhoods of all cells in the result. This took place in the summation depicted in Fig. 6.2 (Chap. 6) but not in Figs. 6.3 and 6.4. Since the cellular c-e structures being added are placed on the identical coordinate system, they are of the same dimension. The domain (set of nodes), i.e. the carrier of cellular c-e structures $U$ and $V$ in Fig. 6.2 is the same, they differ only in the connection arrangement, so the cells in $U$ and $V$ of the same coordinates have different neighbours. In accordance with the axioms of addition, the result $W$ has the same domain, and the neighbourhood of each cell in $W$ is the union of neighbourhoods of this cell in $U$ and $V$, thus neighbourhoods of cells in $W$ are equinumerous. But in Fig. 6.3 only cells with both even coordinates are in $U'$ and with any possible parity in $V$—as in Fig. 6.2. The result $W'$ has the same domain as $V$, and cells with both even coordinates surrounded by their neighbours from $U'$ and $V$, whereas other cells—by their neighbours from $V$ only. Thus the c-e structure $W'$ comprises cells with 8 and with 4 neighbours, so is not cellular. The same reasoning applies to summation shown in Fig. 6.4—replace $V$ with $U$ and $W'$ with $W''$. In symbols, these cases are:

For the case in Fig. 6.2 (Chap. 6):

$$car(U) = car(V), \qquad \underset{U}{nbh}(a_{x,y}) \cap \underset{V}{nbh}(a_{x,y}) = \varnothing,$$

$$\underset{W}{nbh}(a_{x,y}) = \underset{U}{nbh}(a_{x,y}) \cup \underset{V}{nbh}(a_{x,y}),$$

$$\left|\underset{U}{nbh}(a_{x,y})\right| = \left|\underset{V}{nbh}(a_{x,y})\right|, \qquad \left|\underset{W}{nbh}(a_{x,y})\right| = \left|\underset{U}{nbh}(a_{x,y})\right| + \left|\underset{V}{nbh}(a_{x,y})\right|.$$

For the case in Fig. 6.3:

$$car(U') \subsetneq car(V), \qquad \underset{U'}{nbh}(a_{x,y}) \cap \underset{V}{nbh}(a_{x,y}) = \varnothing,$$

$$\underset{W'}{nbh}(a_{x,y}) = \begin{cases} \underset{U'}{nbh}(a_{x,y}) \cup \underset{V}{nbh}(a_{x,y}) & \text{if } a_{x,y} \in car(V) \cap car(U'), \\ \underset{V}{nbh}(a_{x,y}) & \text{if } a_{x,y} \in car(V) \backslash car(U'), \end{cases}$$

$$|\underset{W'}{nbh}(a_{x,y})| = \begin{cases} |\underset{U'}{nbh}(a_{x,y})| + |\underset{V}{nbh}(a_{x,y})| & \text{if } a_{x,y} \in car(V) \cap car(U'), \\ |\underset{V}{nbh}(a_{x,y})| & \text{if } a_{x,y} \in car(V) \backslash car(U'). \end{cases}$$

For the case in Fig. 6.4—replace $V$ with $U$ in Fig. 6.3. Note that addends with different carriers yield not equinumerous neighbourhoods in the result. Remaining cases to check if summation of two cellular c-e structures yields results with atoms equivalent

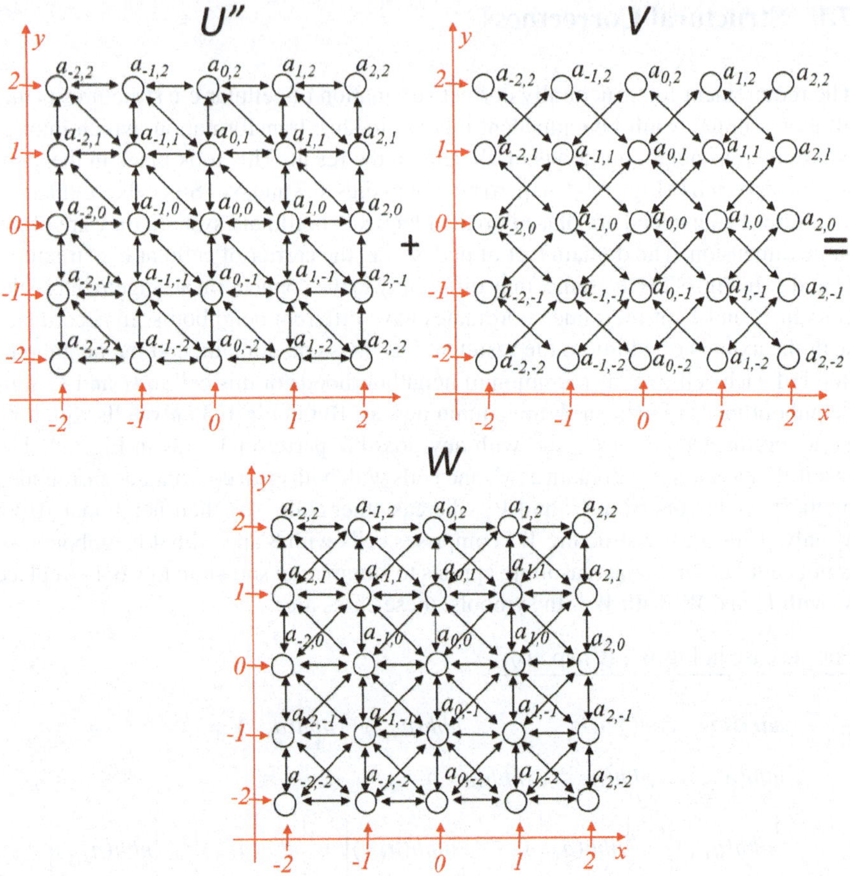


**Fig. 7.2** Fragments of cellular c-e structures $U''$, $V$ and the outcome of summation $W = U'' + V$ being cellular c-e structures with atoms equivalent relative-to shift transformation, thus with equinumerous neighbourhoods of cells

relative-to shift transformation (thus with eqinumerous neighbourhoods) may be illustrated by Figs. 7.2, 7.3.[5]

For the case in Fig. 7.2:

$$car(U'') = car(V), \qquad \underset{U''}{nbh}(a_{x,y}) \cap \underset{V}{nbh}(a_{x,y}) = \{a_{x-1,y+1}, a_{x+1,y-1}\},$$

$$\underset{W}{nbh}(a_{x,y}) = \underset{U''}{nbh}(a_{x,y}) \cup \underset{V}{nbh}(a_{x,y}) =$$

$$\{a_{x-1,y}, a_{x-1,y+1}, a_{x,y+1}, a_{x+1,y}, a_{x+1,y-1}, a_{x,y-1}\} \cup$$

[5] Geometrically, the cells in cellular c-e structure $U''$ may look like hexagons, but with such arrangement on Cartesian plane do not tessellate the plane. Neither the octagonal cells in $W$ do this—see Fig. 2.6 (Chap. 2).

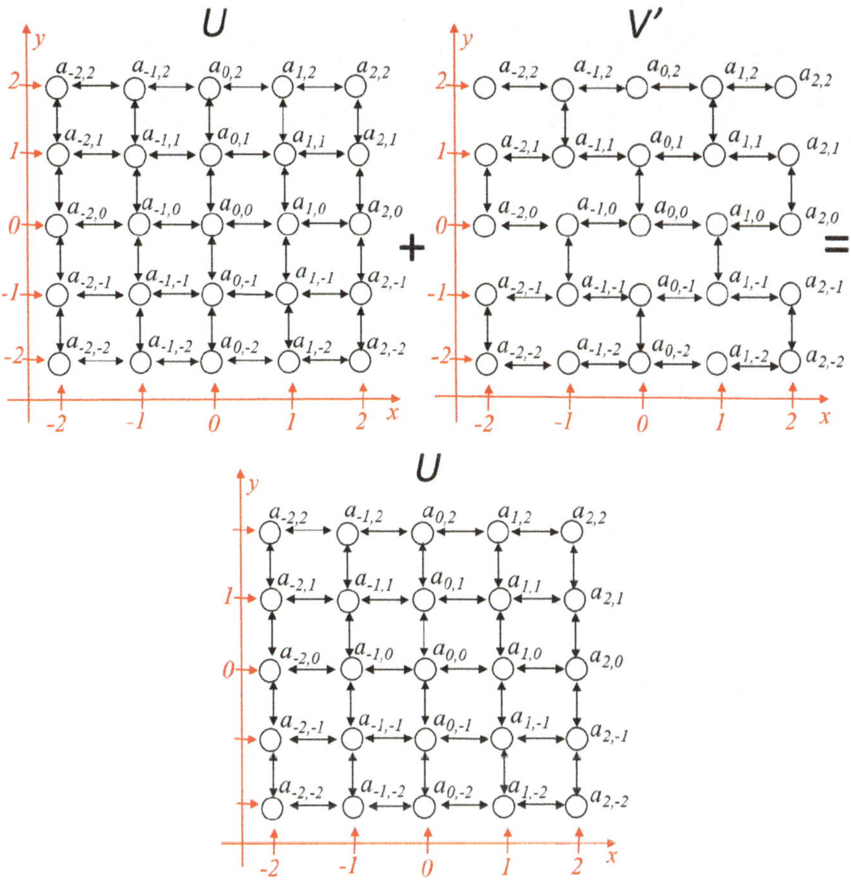


**Fig. 7.3** Fragments of c-e structures $U$, $V'$ and outcome of summation $U = U + V'$. Each has equinumerous neighbourhoods of cells, but only $U$, not $V'$ has all atoms equivalent relative-to the shift transformation. According to cellular automata assumptions, $V'$ is not cellular (recall also Fig. 3.11 in Chap. 3, the structure identical with $V'$)

$$\{a_{x-1,y-1}, a_{x-1,y+1}, a_{x+1,y+1}, a_{x+1,y-1}\} =$$

$$\{a_{x-1,y-1}, a_{x-1,y}, a_{x-1,y+1}, a_{x,y+1}, a_{x+1,y+1}, a_{x+1,y}, a_{x+1,y-1}, a_{x,y-1}\},$$

$$\left|\underset{W}{nbh}(a_{x,y})\right| = \left|\underset{U''}{nbh}(a_{x,y})\right| + \left|\underset{V}{nbh}(a_{x,y})\right| - \left|\underset{U''}{nbh}(a_{x,y}) \cap \underset{V}{nbh}(a_{x,y})\right|.$$

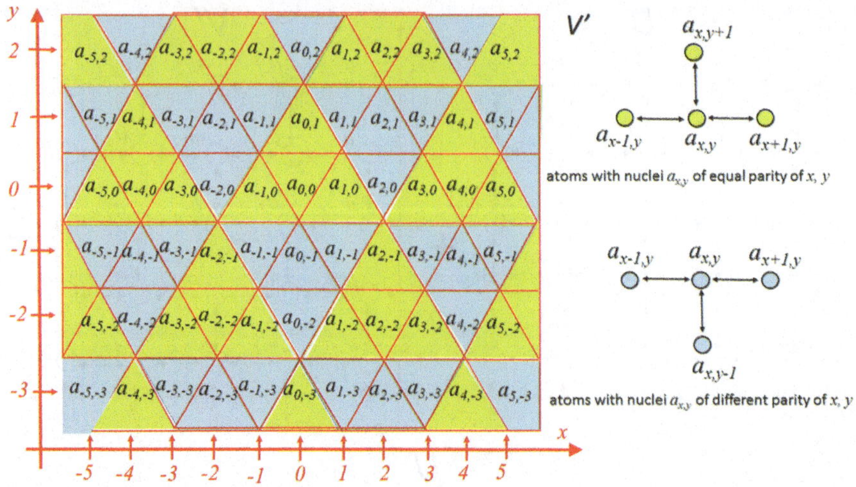


**Fig. 7.4** This c-e structure covers a surface with yellow and blue triangles representing atoms with equal (yellow) and different (blue) parity of nuclei coordinates. Each triangular nucleus shares sides with neighbours of its own colour. It is clearly seen that atoms of the same colour only are equivalent relative-to shift transformation in such arrangement

For the case in Fig. 7.3:

$$car(U) = car(V'), \qquad \underset{V'}{nbh}(a_{x,y}) \subsetneq \underset{U}{nbh}(a_{x,y}),$$

$$\underset{U}{nbh}(a_{x,y}) \setminus \underset{V'}{nbh}(a_{x,y}) = \begin{cases} \{a_{x,y-1}\} \text{ for equal parity of } x, y, \\ \{a_{x,y+1}\} \text{ for different parity of } x, y, \end{cases}$$

$$\underset{U}{nbh}(a_{x,y}) \cup \underset{V'}{nbh}(a_{x,y}) = \underset{U}{nbh}(a_{x,y}).$$

Chapter 3, the structure identical with $V'$).

The c-e structure $V'$ is identical with that in Fig. 3.11 (Chap. 3) which represents a certain triangulation of a surface. Let us picture it in Fig. 7.4, followed by Fig. 7.5, representing different triangulation, thus with different arrangement of atoms, marked yellow and blue in both figures.

The summation of the c-e structures $V''$ and $V'$ depicted in Fig. 7.6 yields the cellular c-e structure $U$.

Here:

$$car(V'') = car(V'), \qquad \underset{V''}{nbh}(a_{x,y}) \cap \underset{V'}{nbh}(a_{x,y}) = \{a_{x-1,y}, a_{x,y}, a_{x+1,y}\},$$

$$\underset{U}{nbh}(a_{x,y}) = \underset{V''}{nbh}(a_{x,y}) \cup \underset{V'}{nbh}(a_{x,y}),$$

$$\left|\underset{V''}{nbh}(a_{x,y})\right| = \left|\underset{V'}{nbh}(a_{x,y})\right| = 3, \quad \left|\underset{U}{nbh}(a_{x,y})\right| = \left|\underset{V''}{nbh}(a_{x,y})\right| + \left|\underset{V'}{nbh}(a_{x,y})\right| - 2 = 4.$$

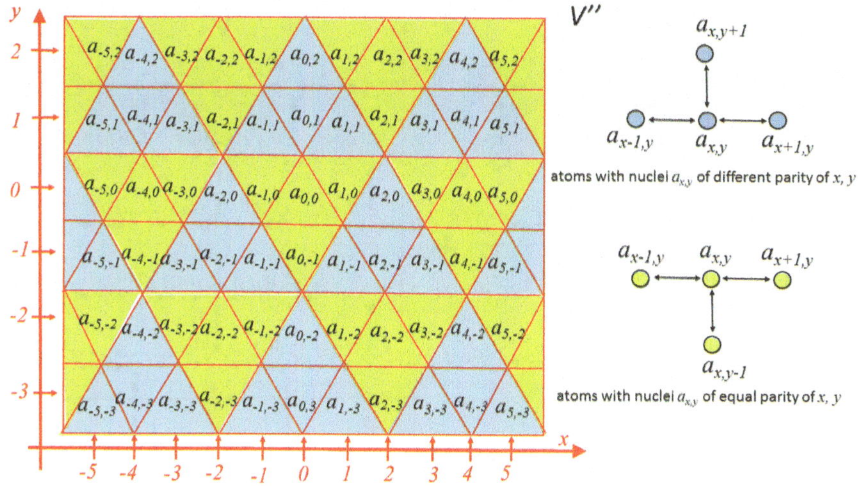


**Fig. 7.5** Different covering of a surface than that in Fig. 7.4, but with the same comment

Obviously, cellular c-e structures with equal carriers and equal neighbourhoods of the shared cells are structurally equal—like those in Fig. 5.2 (Chap. 5). Their sum $U_{red} + U_{blue} + U_{green}$
depicted in Fig. 6.6 (Chap. 6) is structurally the same as components, but may be the result of summation of different components, for instance $Z = X + Y$ depicted in Fig. 7.7, where:

$$car(X) = car(Y), \qquad \underset{X}{nbh}(a_{x,y}) \cap \underset{Y}{nbh}(a_{x,y}) = \varnothing,$$

$$\underset{Z}{nbh}(a_{x,y}) = \underset{X}{nbh}(a_{x,y}) \cup \underset{Y}{nbh}(a_{x,y}),$$

$$\left|\underset{X}{nbh}(a_{x,y})\right| = \left|\underset{Y}{nbh}(a_{x,y})\right| = 3, \quad \left|\underset{Z}{nbh}(a_{x,y})\right| = \left|\underset{X}{nbh}(a_{x,y})\right| + \left|\underset{Y}{nbh}(a_{x,y})\right| = 6.$$

Modifying disconnected cellular c-e structure $X$ to connected $X'$ and summing it with $Y$, one obtains the same cellular c-e structure $Z$ (Fig. 7.8).

Here:

$$car(X') = car(Y),$$

$$\underset{X'}{nbh}(a_{x,y}) \cap \underset{Y}{nbh}(a_{x,y}) = \begin{cases} \{a_{x,y-2}\} \text{ for } a_{x,y} \text{ on white diagonals,} \\ \{a_{x,y+2}\} \text{ for } a_{x,y} \text{ on grey diagonals,} \end{cases}$$

$$\underset{Z}{nbh}(a_{x,y}) = \underset{X'}{nbh}(a_{x,y}) \cup \underset{Y}{nbh}(a_{x,y}),$$

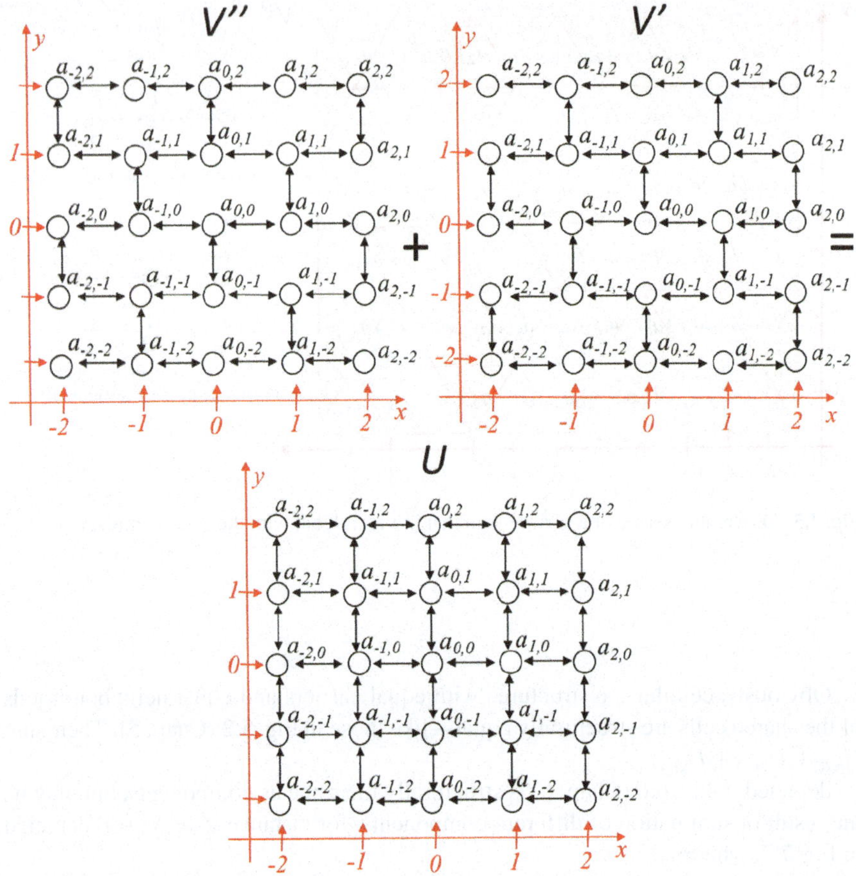

**Fig. 7.6** Fragments of c-e structures $V''$, $V'$ and their sum $U = V'' + V'$ depicted in Fig. 7.3

$$\left|\underset{X'}{nbh}(a_{x,y})\right| = 4, \quad \left|\underset{Y}{nbh}(a_{x,y})\right| = 3, \quad \left|\underset{Z}{nbh}(a_{x,y})\right| = 6.$$

The above and the next examples illustrate an observation that equality of carriers of addends, unless disjoint, is needed for correct result of addition. Indeed, if only one component of addition hosts a cell $a_{x,y}$ then this cell in the result has the same neighbourhood as in this component (remember that neighbourhoods in the result of addition are unions of respective neighbourhoods in the addends). See for instance Figs. 6.3, 6.4 (Chap. 6). But equality of carriers is not sufficient condition for correct result of summing up cellular c-e structures, as shows Fig. 7.9.

To verify whether neighbourhoods of all cells in the result of summation are equinumerous, first add atoms with the same nuclei in the addends, then sum up all atoms resulted in this way. If the outcome of the latter summation fulfils this

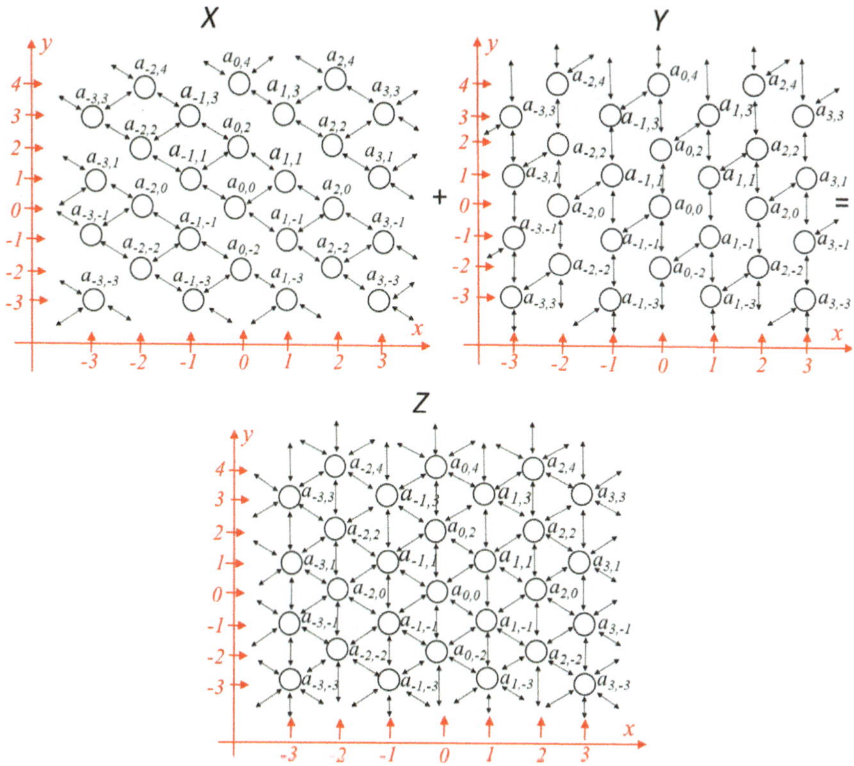


**Fig. 7.7** The cellular c-e structure $Z = X + Y$ structurally the same as $U_{rbg} = U_{red} + U_{blue} + U_{green}$ in Fig. 6.6 (Chap. 6) and that in Fig. 2.9 (Chap. 2). The component $X$ is the disconnected cellular c-e structure, which from the cellular automata perspective is the infinite family of connected cellular automata (bounded), in contrast to the concept of standard c-e structures, admitting disconnectedness

requirement, the finally obtained c-e structure is structurally correct. Otherwise is not. This is insufficient requirement for semantic correctness, which will be analysed soon. Similar reasoning applies to cellular c-e structures of dimension higher than 2. These considerations on structural feasibility of combining c-e structures into cellular, bring to the conclusion:

*The result of summation of c-e structures is structurally correct if and only if carriers of the components being summed up are identical and above described verification of the equal size of cells' neighbourhoods, provides positive answer.*[a]

---

[a] Yet mind about summation of non-cellular c-e structures with different carriers, to place the resulted cellular one onto cylinders or toruses.

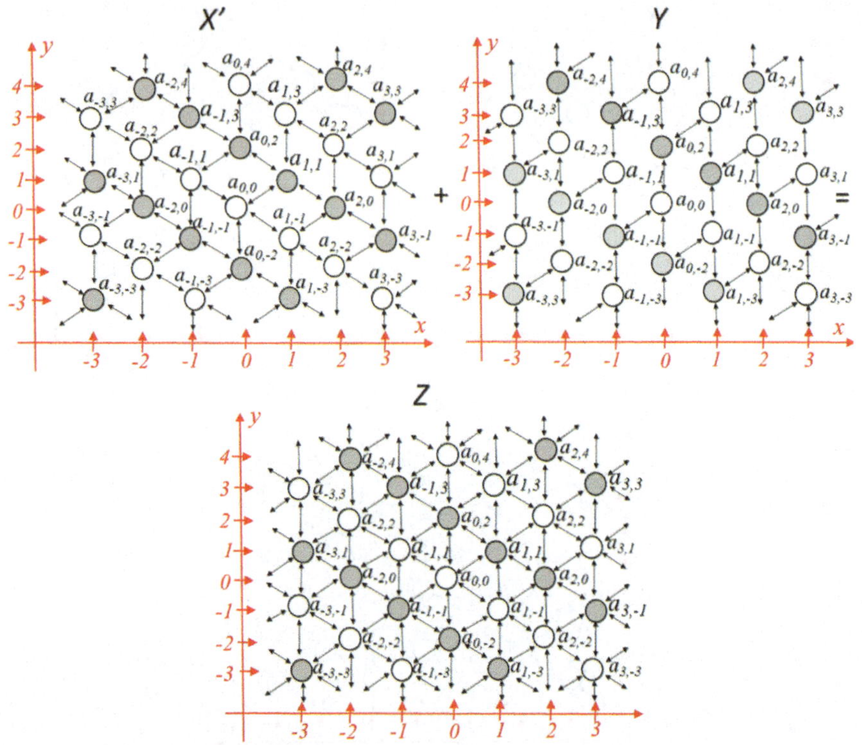


**Fig. 7.8** The cellular c-e structure $Z = X' + Y$ obtained from the connected component $X'$, a modification of disconnected $X$ in Fig. 7.7

## 7.2 Semantical Correctness

The requirement for semantically correct summation $W = U + V$ of cellular c-e structures is that each table $B_{\mathbf{j}\to\mathbf{i}}$ in the result $W$ contains functions $f_k^{\mathbf{j}\to\mathbf{i}}$ (mappings of each argument onto a single value) unambiguously determining change value of a cell $a_{x,y}$ from **j** to **i**. So, there are no conflicts on entries of resulting tables and between functions in different tables—see, for instance, Tables 6.5, 6.6, 6.7 (Chap. 6). This requirement is evident: each cell in the outcome of summation should assume well determined single status (value) in a given state. Remember that the patterns $f_k^{\mathbf{j}\to\mathbf{i}}$ are represented by rows of tables $B_{\mathbf{j}\to\mathbf{i}}$ and each entry in the row corresponds to a neighbour of the cell $a_{x,y}$ whose status is determined by means of semantic formulas containing appropriate tables. Note that the neighbourhood of a cell $a_{x,y}$ in the result of summation is always a union of neighbourhoods of $a_{x,y}$ for $a_{x,y}$ belonging to all components. Thus, the top line $nbh(a_{x,y})$ in the tables $B_{\mathbf{j}\to\mathbf{i}}$ of the result (the domain of functions $f_k^{\mathbf{j}\to\mathbf{i}}$), contains this union, i.e. a set of arguments of each $f_k^{\mathbf{j}\to\mathbf{i}}$. To ensure correct and intended (for a modelled task) result of summation,

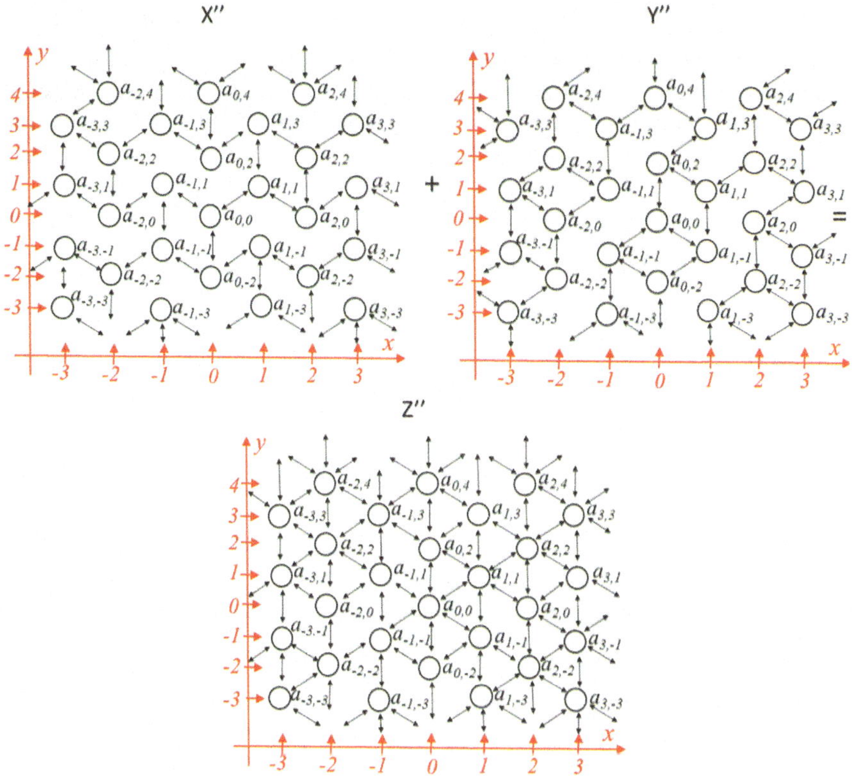


**Fig. 7.9** The same carriers in $X''$ and $Y''$ and $Z''$ but some cells in $Z'' = X'' + Y''$ have 6 neighbours (e.g. $a_{0,0}$), while some others only 5 (e.g. $a_{-1,1}$) and 4 (e.g. $a_{0,-2}$). So, cellular addends yield here non-cellular result

mutual location dependencies (set-theoretic) between neighbourhoods of cells in the components $U$, $V$ have to be taken into regard, such as: $\underset{U}{nbh}(a_{x,y}) \cap \underset{V}{nbh}(a_{x,y}) = \varnothing$, $\underset{U}{nbh}(a_{x,y}) \cap \underset{V}{nbh}(a_{x,y}) \neq \varnothing$, $\underset{U}{nbh}(a_{x,y}) \subsetneq \underset{V}{nbh}(a_{x,y})$, $\underset{U}{nbh}(a_{x,y}) = \underset{V}{nbh}\,(a_{x,y})$, as well as a type of semantics involved. One type stipulates that behaviour of a cell $a_{x,y}$ is determined only by its value and a number of neighbours assuming certain values (like in the Game of life), let us call it "Counting **V**alues of **N**eighbours", **CVN** in short. The other type stipulates that behaviour of a cell $a_{x,y}$ is determined by value of $a_{x,y}$ and exact distribution, i.e. placement (inside neighbourhood) of neighbours holding certain values (like in example of light rays). Let us call it "**D**istribution of **V**alues of **N**eighbours", **DVN** in short. So, construction of tables $B_{\mathrm{j}\to\mathrm{i}}$ depends on the characteristics CVN or DVN of cellular c-e structures in question, in structural as well as semantic case. Let us introduce a short notation for semantic tables $B_{\mathrm{j}\to\mathrm{i}}$, (to avoid writing out them on the whole) and begin with examples of type CVN.

The tables $B^{birth}_{\mathbf{0}\to\mathbf{1}}$, $B^{surv}_{\mathbf{1}\to\mathbf{1}}$ (Tables 5.1 5.2 in Chap. 5) for behaviour of a cell $a_{x,y}$ in the von Neumann type Game of life, may be denoted as $B[4]^{(3*\mathbf{1})}_{\mathbf{0}\to\mathbf{1}}$, $B[4]^{(3*\mathbf{1})\veebar(2*\mathbf{1})}_{\mathbf{1}\to\mathbf{1}}$ respectively, where "4" is number of neighbours of $a_{x,y}$, (3*$\mathbf{1}$) means "3 neighbours assume value $\mathbf{1}$", whereas (3*$\mathbf{1}$) $\veebar$ (2*$\mathbf{1}$) means "3 or 2 neighbours assume value $\mathbf{1}$" and "$\mathbf{j} \to \mathbf{i}$"—as always: change of value of $a_{x,y}$ from $\mathbf{j}$ to $\mathbf{i}$. The tables for the Moore type Game of life: replace only "4" with "8". If the name of c-e structure (cellular or not) involved is $U$, then let write $\underset{U}{B}[4]^{(3*\mathbf{1})}_{\mathbf{0}\to\mathbf{1}}$, etc. The summation $W = U + V$ exemplified in Fig. 6.2 (Chap. 6) effects in construction $\underset{W}{B}[8]^{(3*\mathbf{1})}_{\mathbf{0}\to\mathbf{1}}$ and $\underset{W}{B}[8]^{(3*\mathbf{1})\veebar(2*\mathbf{1})}_{\mathbf{1}\to\mathbf{1}}$, shown explicitly as Tables 6.5, 6.6. Let us denote corresponding composition of semantic tables for behaviour of $a_{x,y}$ by $\underset{W}{B}[8]^{(3*\mathbf{1})}_{\mathbf{0}\to\mathbf{1}} = \underset{U}{B}[4]^{(3*\mathbf{1})}_{\mathbf{0}\to\mathbf{1}} \oplus \underset{V}{B}[4]^{(3*\mathbf{1})}_{\mathbf{0}\to\mathbf{1}}$ and $\underset{W}{B}[8]^{(3*\mathbf{1})\veebar(2*\mathbf{1})}_{\mathbf{1}\to\mathbf{1}} = \underset{U}{B}[4]^{(3*\mathbf{1})\veebar(2*\mathbf{1})}_{\mathbf{1}\to\mathbf{1}} \oplus \underset{V}{B}[4]^{(3*\mathbf{1})\veebar(2*\mathbf{1})}_{\mathbf{1}\to\mathbf{1}}$.

So, operation "$\oplus$" scatters 3 values $\mathbf{1}$ in each of $\binom{8}{3} = 56$ rows of the 56-row table $\underset{W}{B}[8]^{(3*\mathbf{1})}_{\mathbf{0}\to\mathbf{1}}$, as well as 3 values $\mathbf{1}$ in $\binom{8}{3} = 56$ rows of table $\underset{W}{B}[8]^{(3*\mathbf{1})\veebar(2*\mathbf{1})}_{\mathbf{1}\to\mathbf{1}}$ along with $\binom{8}{2} = 28$ remaining rows of this 84-row table. The arrangement of neighbours around each cell $a_{x,y}$ shared by $U$ and $V$ is such that $\underset{U}{nbh}(a_{x,y}) \cap \underset{V}{nbh}(a_{x,y}) = \varnothing$ and $\underset{U}{nbh}(a_{x,y}) \cup \underset{V}{nbh}(a_{x,y}) = \underset{W}{nbh}(a_{x,y})$ (see Figs. 6.1, 6.2 and Tables 6.1, 6.2, 6.3 and 6.4, in Chap. 6), thus compositions $\underset{U}{B}[4]^{(3*\mathbf{1})}_{\mathbf{0}\to\mathbf{1}} \oplus \underset{V}{B}[4]^{(3*\mathbf{1})}_{\mathbf{0}\to\mathbf{1}}$ and $\underset{U}{B}[4]^{(3*\mathbf{1})\veebar(2*\mathbf{1})}_{\mathbf{1}\to\mathbf{1}} \oplus \underset{V}{B}[4]^{(3*\mathbf{1})\veebar(2*\mathbf{1})}_{\mathbf{1}\to\mathbf{1}}$ do not create conflicts on entries of tables $\underset{W}{B}[8]^{(3*\mathbf{1})}_{\mathbf{0}\to\mathbf{1}}$ and $\underset{W}{B}[8]^{(3*\mathbf{1})\veebar(2*\mathbf{1})}_{\mathbf{1}\to\mathbf{1}}$. So, every function $f_k''^{\mathbf{0}\to\mathbf{1}}$ in $\underset{W}{B}[8]^{(3*\mathbf{1})}_{\mathbf{0}\to\mathbf{1}}$ and $f_k''^{\mathbf{1}\to\mathbf{1}}$ in $\underset{W}{B}[8]^{(3*\mathbf{1})\veebar(2*\mathbf{1})}_{\mathbf{1}\to\mathbf{1}}$ maps its argument onto a single value.

Anticipating formulation of the CVN type behaviour of many-valued cells, note that semantic tables should specify also a number of values $\mathbf{0}$ in rows, so let us write $\underset{W}{B}[8]^{(3*\mathbf{1},\,5*\mathbf{0})}_{\mathbf{0}\to\mathbf{1}}$ and $\underset{W}{B}[8]^{(3*\mathbf{1},\,5*\mathbf{0})\veebar(2*\mathbf{1},\,6*\mathbf{0})}_{\mathbf{1}\to\mathbf{1}}$ instead of $\underset{W}{B}[8]^{(3*\mathbf{1})}_{\mathbf{0}\to\mathbf{1}}$ and $\underset{W}{B}[8]^{(3*\mathbf{1})\veebar(2*\mathbf{1})}_{\mathbf{1}\to\mathbf{1}}$. So, e.g. the superscript in $\underset{W}{B}[8]^{(3*\mathbf{1},\,5*\mathbf{0})\veebar(2*\mathbf{1},\,6*\mathbf{0})}_{\mathbf{1}\to\mathbf{1}}$ is read up as "56 rows of the Table 6.6 (Chap. 6) comprise three values $\mathbf{1}$ and five values $\mathbf{0}$, besides 28 remaining rows comprise two values $\mathbf{1}$ and six values $\mathbf{0}$". This prompts a general notational form of the CVN type semantic table for a cell $a_{x,y}$ in cellular c-e structures with size $m$ of the cells' neighbourhood:

$$B[m]_{\mathbf{j}\to\mathbf{i}}^{\overbrace{(\lambda_{11}*\mathbf{q}_{11},\ \lambda_{12}*\mathbf{q}_{12},\ \ldots,\ \lambda_{1h_1}*\mathbf{q}_{1h_1})}^{\text{group 1}}\veebar\overbrace{(\lambda_{21}*\mathbf{q}_{21},\ \lambda_{22}*\mathbf{q}_{22},\ \ldots,\ \lambda_{2h_2}*\mathbf{q}_{2h_2})}^{\text{group 2}},\ldots}$$

where $\lambda_{kl}$ are natural numbers—coefficients[6] (multiples) of values $\mathbf{q}_{kl}$ of neighbours of $a_{x,y}$. Each group in the superscript represents a set of rows (i.e. functions $f_k^{\mathbf{j}\to\mathbf{i}}$), and every row in the group comprises $\lambda_{kl}$ of values $\mathbf{q}_{kl}$. Thus the coefficients should satisfy:

$$\begin{aligned}
\lambda_{11} + \lambda_{12} + \ldots + \lambda_{1h_1} &= m \\
\lambda_{21} + \lambda_{22} + \ldots + \lambda_{2h_2} &= m \\
\ldots\ldots\ldots\ldots\ldots\ldots\ldots\ldots\ldots\ldots \\
\lambda_{r1} + \lambda_{r2} + \ldots + \lambda_{rh_r} &= m
\end{aligned}$$

where $m = |nbh(a_{x,y})|$, the size of cells' neighbourhood in a cellular c-e structure in question.

This means that the table $B[m]^{\cdots}_{\mathbf{j}\to\mathbf{i}}$ contains rows with:

$$\begin{array}{ll}
\lambda_{11} \text{ values } \mathbf{q}_{11},\ \lambda_{12} \text{ values } \mathbf{q}_{12},\ \ldots,\ \lambda_{1h_1} \text{ values } \mathbf{q}_{1h_1} & \text{(group 1)} \\
\lambda_{21} \text{ values } \mathbf{q}_{21},\ \lambda_{22} \text{ values } \mathbf{q}_{22},\ \ldots,\ \lambda_{2h_1} \text{ values } \mathbf{q}_{2h_1} & \text{(group 2)} \\
\ldots\ldots\ldots\ldots\ldots\ldots\ldots\ldots\ldots\ldots\ldots\ldots\ldots\ldots\ldots\ldots\ldots\ldots\ldots\ldots & \\
\lambda_{r1} \text{ values } \mathbf{q}_{r1},\ \lambda_{r2} \text{ values } \mathbf{q}_{r2},\ \ldots,\ \lambda_{rh_r} \text{ values } \mathbf{q}_{rh_r} & \text{(group } r\text{)}
\end{array}$$

where $h_1, h_2, \ldots, h_r$ are sizes of the groups. In above example:

$$m = 8,\ \mathbf{j} = \mathbf{0},\ \mathbf{i} = \mathbf{1},\ \lambda_{11} = 3,\ \mathbf{q}_{11} = \mathbf{1},\ \lambda_{12} = 5,\ \mathbf{q}_{12} = \mathbf{0},$$

for the table $\underset{W}{B}[8]^{\overbrace{(3*\mathbf{1},\ 5*\mathbf{0})}^{\text{group 1}}}_{\mathbf{0}\to\mathbf{1}}$ and

$$m = 8,\ \mathbf{j} = \mathbf{1},\ \mathbf{i} = \mathbf{1},$$
$$\lambda_{21} = 3,\ \mathbf{q}_{21} = \mathbf{1},\ \lambda_{22} = 5,\ \mathbf{q}_{22} = \mathbf{0},\ \lambda_{23} = 2,\ \mathbf{q}_{23} = \mathbf{1},\ \lambda_{24} = 6,\ \mathbf{q}_{24} = \mathbf{0},$$

for the table $\underset{W}{B}[8]^{\overbrace{(3*\mathbf{1},\ 5*\mathbf{0})}^{\text{group 1}}\veebar\overbrace{(2*\mathbf{1},\ 6*\mathbf{0})}^{\text{group 2}}}_{\mathbf{1}\to\mathbf{1}}$.

In this example, the next-state rule encoded in tables $\underset{W}{B}[8]^{\cdots}_{\cdots}$ for $W$ (Moore type Game of life) has been the same as for $U$ and $V$ (von Neumann type)—tables $\underset{U}{B}[4]^{\cdots}_{\cdots}$, $\underset{V}{B}[4]^{\cdots}_{\cdots}$. This was the intended decision of the task inventor(s), not the consequence of principles of cellular automata and cellular c-e structures. As mentioned before,

---

[6] The number $\lambda_{kl}$ of cells with value $\mathbf{q}_{kl}$ in the neighbourhood of the cell $a_{x,y}$ is called sometimes a *repetition coefficient of* $\mathbf{q}_{kl}$, and a group—a *set with repetitions*. If a value does not occur in a group, then its repetition coefficient is 0. Remember that values of cells are here denoted by boldface numerals, so the symbols $\mathbf{q}_{kl}$ represent boldface numerals.

the next-state rule is specific to a cellular c-e structure being devised, not common to all of them, as in case of standard c-e structures presented in Chap. 1.

After these considerations on the general form of semantic tables in case of CVN type semantics, it is seen that no conflict on entries occurs in tables constructed out of tables for cellular c-e structures being components of summation, when neighbourhoods of cells in different components are pairwise disjoint. Indeed: the neighbourhoods are domains of functions $f_k^{\mathbf{j}\to\mathbf{i}}$ that come from different summation components and have ranges being different rows of respective table. Thus, disjointness of the neighbourhoods in different components suffices for correct summation. But this is not the necessary condition: it may happen that summation of $U$, $V$ such that $\underset{U}{nbh}(a_{x,y}) \cap \underset{V}{nbh}(a_{x,y}) \neq \varnothing$, would not yield conflict on a certain entry in the resulted table. This takes place when for a certain $e \in \underset{U}{nbh}(a_{x,y}) \cap \underset{V}{nbh}(a_{x,y})$, respective functions $f_k^{\mathbf{j}\to\mathbf{i}}$ in tables for $U$ and $V$ return the same value for argument $e$. Obviously, if this takes place for all such cells $e$ then correct summation is guaranteed too, provided that all cells in $U + V$ have equinumerous neighbourhood – in case of the CVN semantics, or all atoms in $U + V$ are equivalent relative-to shift – in case of the DVN semantics (then all cells have equinumerous neighbourhood too). So, the disjointness of neighbourhoods in $U$ and $V$ is overmuch strong condition for correct summation.

Notice that the values $\mathbf{q}_{kl}$ and their repetition coefficients $\lambda_{kl}$ in the table $B[m]_{\mathbf{j}\to\mathbf{i}}$ obtained in effect of summation of cellular c-e structures, are so chosen as to define intended (in the task modelled) behaviour of a cell $a_{x,y}$. That is the change its value (status) from $\mathbf{j}$ to $\mathbf{i}$. Notice also that the requirement for the equivalence of atoms relative-to shift in case of the CVN type semantics is superfluous. It matters in case of the DVN type semantics.

As an example of correct summation in case of the DVN semantics, let us take the tables $B^{red}_{\mathbf{0}\to\mathbf{1}}$, $B^{blue}_{\mathbf{0}\to\mathbf{1}}$, $B^{green}_{\mathbf{0}\to\mathbf{1}}$ (Tables 5.3, 5.4, 5.5 in Chap. 5) specifying intended behaviour of a cell $a_{x,y}$ in the cellular c-e structures $U_{red}$, $U_{blue}$, $U_{green}$ (Fig. 5.2) and their sum $U_{rbg}$ (Fig. 6.6 in Chap. 6). The semantic table $B^{rbg}_{\mathbf{0}\to\mathbf{1}}$ (Table 6.7) specifying intended behaviour of $U_{rbg} = U_{red} + U_{blue} + U_{green}$ is obtained from the component tables by their "downward concatenation", denoted by $B^{rbg}_{\mathbf{0}\to\mathbf{1}} = B^{red}_{\mathbf{0}\to\mathbf{1}} \Cup B^{blue}_{\mathbf{0}\to\mathbf{1}} \Cup B^{green}_{\mathbf{0}\to\mathbf{1}}$. Collection of these subtables into integrated Table 6.7 emphasizes simultaneous, not separate, evolution of $U_{red}$, $U_{blue}$, $U_{green}$, depicted in Fig. 6.7 and determines behaviour of any cell $a_{x,y}$ inside their sum $U_{rbg}$ in Fig. 6.6. So works the multigraph-automaton in Fig. 7.1.

Another example presents summation of c-e structures $V''$ and $V'$ shown in Fig. 7.6, with equinumerous neighbourhoods of cells, but each with not equivalent (relative-to shift) atoms. Their cells assume values $\mathbf{0}$ ("empty") and $\mathbf{1}$ ("full") only. Suppose the DVN semantics and such behaviour: if two neighbouring cells have different values in a state, then they change their values into inverse in the next state. This is specified by tables $B^{V''}_{\mathbf{j}\to\mathbf{i}}$, $B^{V'}_{\mathbf{j}\to\mathbf{i}}$, $B^{U}_{\mathbf{j}\to\mathbf{i}}$ $\mathbf{j}, \mathbf{i} \in \{\mathbf{0},\mathbf{1}\}, \mathbf{j} \neq \mathbf{i}$ (Tables 7.1, 7.2, 7.3). Relation $[\![a_{x,y}]\!]$ determining change of value of $a_{x,y}$ in c-e structures $V''$, $V'$ and $U$ from $\mathbf{j}$ in a state $s$ to $\mathbf{i}$ in the next state $t$ is defined as:

**Table 7.1** $B^{V''}_{\mathbf{j}\to\mathbf{i}}$. Entries contain values of functions $f_k^{''\mathbf{j}\to\mathbf{i}} \in B^{V''}_{\mathbf{j}\to\mathbf{i}}$ with domain $\underset{V''}{nbh}(a_{x,y})$

| $\underset{V''}{nbh}(a_{x,y})$ | $a_{x-1,y}$ | $a_{x,y-1}$ | $a_{x+1,y}$ |
|---|---|---|---|
| $f_1^{''\mathbf{j}\to\mathbf{i}}$ | **j** | **j** | **i** |
| $f_2^{''\mathbf{j}\to\mathbf{i}}$ | **j** | **i** | **j** |
| $f_3^{''\mathbf{j}\to\mathbf{i}}$ | **i** | **j** | **j** |
| $f_4^{''\mathbf{j}\to\mathbf{i}}$ | **j** | **i** | **i** |
| $f_5^{''\mathbf{j}\to\mathbf{i}}$ | **i** | **j** | **i** |
| $f_6^{''\mathbf{j}\to\mathbf{i}}$ | **i** | **i** | **j** |
| $f_7^{''\mathbf{j}\to\mathbf{i}}$ | **i** | **i** | **i** |

**Table 7.2** $B^{V'}_{\mathbf{j}\to\mathbf{i}}$.Entries contain values of functions $f_k^{'\mathbf{j}\to\mathbf{i}} \in B^{V'}_{\mathbf{j}\to\mathbf{i}}$ with domain $\underset{V'}{nbh}(a_{x,y})$

| $\underset{V'}{nbh}(a_{x,y})$ | $a_{x-1,y}$ | $a_{x,y+1}$ | $a_{x+1,y}$ |
|---|---|---|---|
| $f_1^{'\mathbf{j}\to\mathbf{i}}$ | **j** | **j** | **i** |
| $f_2^{'\mathbf{j}\to\mathbf{i}}$ | **j** | **i** | **j** |
| $f_3^{'\mathbf{j}\to\mathbf{i}}$ | **i** | **j** | **j** |
| $f_4^{'\mathbf{j}\to\mathbf{i}}$ | **j** | **i** | **i** |
| $f_5^{'\mathbf{j}\to\mathbf{i}}$ | **i** | **j** | **i** |
| $f_6^{'\mathbf{j}\to\mathbf{i}}$ | **i** | **i** | **j** |
| $f_7^{'\mathbf{j}\to\mathbf{i}}$ | **i** | **i** | **i** |

$(s, t) \in [\![a_{x,y}]\!]$ holds iff:

$$\bigvee_{\mathbf{i}\in\{\mathbf{0},\mathbf{1}\}} \left( \left[ \bigvee_{\mathbf{j}\in\{\mathbf{0},\mathbf{1}\}} s(a_{x,y}) = \mathbf{j} \wedge s|_{nbh(a_{x,y})} \subset B^{V''}_{\mathbf{j}\to\mathbf{i}} \right] \wedge t(a_{x,y}) = \mathbf{i} \right)$$

$$\bigvee_{\mathbf{i}\in\{\mathbf{0},\mathbf{1}\}} \left( \left[ \bigvee_{\mathbf{j}\in\{\mathbf{0},\mathbf{1}\}} s(a_{x,y}) = \mathbf{j} \wedge s|_{nbh(a_{x,y})} \in B^{V'}_{\mathbf{j}\to\mathbf{i}} \right] \wedge t(a_{x,y}) = \mathbf{i} \right)$$

$$\bigvee_{\mathbf{i}\in\{\mathbf{0},\mathbf{1}\}} \left( \left[ \bigvee_{\mathbf{j}\in\{\mathbf{0},\mathbf{1}\}} s(a_{x,y}) = \mathbf{j} \wedge s|_{nbh(a_{x,y})} \subset B^{U}_{\mathbf{j}\to\mathbf{i}} \right] \wedge t(a_{x,y}) = \mathbf{i} \right)$$

This presents progressive—to infinity—alternated exchange of values $\mathbf{0}, \mathbf{1}$ between neighbouring cells, starting from value $\mathbf{1}$ of $a_{0,0}$ (token in it). Figure 7.10 depicts six first stages of this exchange and Fig. 7.11 the multigraph-automaton of this task.

Considerations on behavioural peculiarities of cellular c-e structures lead to the following recapitulation. Semantic tables $B_{\mathbf{j}\to\mathbf{i}}$ should be constructed in such way that any formula

**Table 7.3** $B^U_{\mathbf{j}\to\mathbf{i}}$. Entries contain values of functions $f_k^{\prime\prime\prime\mathbf{j}\to\mathbf{i}} \in B^U_{\mathbf{j}\to\mathbf{i}}$ with domain $\underset{U}{nbh}(a_{x,y})$, where $\mathbf{j} \in \{\mathbf{0}, \mathbf{1}\}, \mathbf{i} \in \{\mathbf{0}, \mathbf{1}\}, \mathbf{j} \neq \mathbf{i}$

| $\underset{U}{nbh}(a_{x,y})$ | $a_{x-1,y}$ | $a_{x,y-1}$ | $a_{x+1,y}$ | $a_{x,y+1}$ |
|---|---|---|---|---|
| $f_1^{\prime\prime\prime\mathbf{j}\to\mathbf{i}}$ | **j** | **j** | **i** | **j** |
| $f_2^{\prime\prime\prime\mathbf{j}\to\mathbf{i}}$ | **j** | **i** | **j** | **j** |
| $f_3^{\prime\prime\prime\mathbf{j}\to\mathbf{i}}$ | **i** | **j** | **j** | **j** |
| $f_4^{\prime\prime\prime\mathbf{j}\to\mathbf{i}}$ | **j** | **i** | **i** | **j** |
| $f_5^{\prime\prime\prime\mathbf{j}\to\mathbf{i}}$ | **i** | **j** | **i** | **j** |
| $f_6^{\prime\prime\prime\mathbf{j}\to\mathbf{i}}$ | **i** | **i** | **j** | **j** |
| $f_7^{\prime\prime\prime\mathbf{j}\to\mathbf{i}}$ | **i** | **i** | **i** | **j** |
| $f_8^{\prime\prime\prime\mathbf{j}\to\mathbf{i}}$ | **j** | **j** | **i** | **i** |
| $f_9^{\prime\prime\prime\mathbf{j}\to\mathbf{i}}$ | **j** | **i** | **j** | **i** |
| $f_{10}^{\prime\prime\prime\mathbf{j}\to\mathbf{i}}$ | **i** | **j** | **j** | **i** |
| $f_{11}^{\prime\prime\prime\mathbf{j}\to\mathbf{i}}$ | **j** | **i** | **i** | **i** |
| $f_{12}^{\prime\prime\prime\mathbf{j}\to\mathbf{i}}$ | **i** | **j** | **i** | **i** |
| $f_{13}^{\prime\prime\prime\mathbf{j}\to\mathbf{i}}$ | **i** | **i** | **j** | **i** |
| $f_{14}^{\prime\prime\prime\mathbf{j}\to\mathbf{i}}$ | **i** | **i** | **i** | **i** |
| $f_{15}^{\prime\prime\prime\mathbf{j}\to\mathbf{i}}$ | **j** | **j** | **j** | **i** |

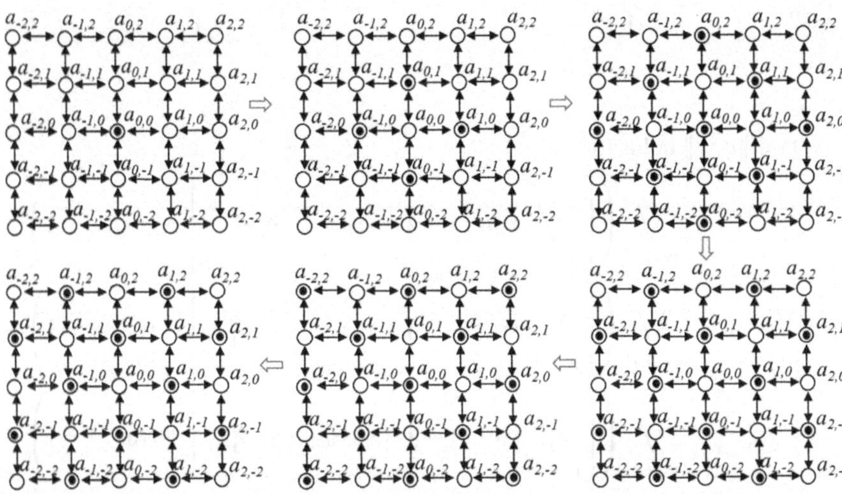

**Fig. 7.10** Six successive stages of cellular c-e structure $U = V'' + V'$ (Fig. 7.6) evolution in CVN semantics specified by Table 7.3. The evolution starts from initial state $s$ with $s(a_{0,0}) = \mathbf{1}$ and $s(a_{x,y}) = \mathbf{0}$ for remaining $a_{x,y}$. The tokens represent value **1**, their lack – value **0**

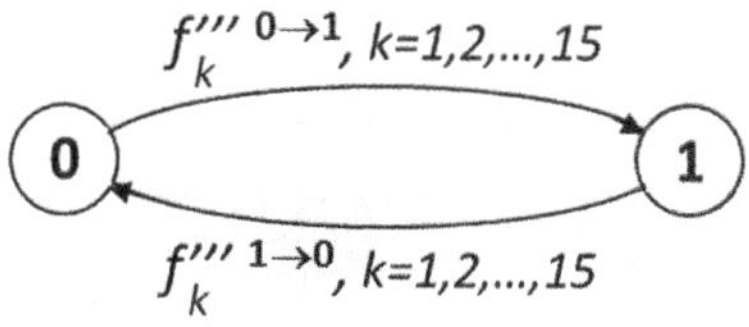


**Fig. 7.11** A multigraph-automaton evoking infinite process of alternated change value of neighbouring cells $a_{x,y}$ in successive stages of activity of $U$. The edge is being traversed from **j** to **i** (**j**, **i** $\in \{0,1\}$, **j** $\neq$ **i**) if restricted state $s|_{a_{x,y}}$ matches certain function $f_k'''^{\mathbf{j}\to\mathbf{i}} \in B^U_{\mathbf{j}\to\mathbf{i}}$ when $U$ is in the state $s$. So, the state of this automaton is the value of such function (pattern) in parameter $B^U_{\mathbf{j}\to\mathbf{i}}$, not the state of cellular c-e structure $U$. [10]

$$\bigvee_{\mathbf{i}\in\mathbf{Q}}\left(\left[\bigvee_{\mathbf{j}\in\mathbf{Q}} s(a_{x.y,\ldots,z}) = \mathbf{j} \wedge s|_{nbh(a_{x.y,\ldots,z})} \in B_{\mathbf{j}\to\mathbf{i}}\right] \wedge t(a_{x.y,\ldots,z}) = \mathbf{i}\right)$$

which expresses change value of the cell $a_{x.y,\ldots,z}$ from $\mathbf{j} \in Q$ in state $s$ into $\mathbf{i} \in Q$ in state $t$, thus defines relation $(s,t) \in [\![a_{x.y,\ldots,z}]\!]$, ensures this change unambiguously. This means that cannot exist function (pattern) in tables $B_{\mathbf{j}\to\mathbf{i}}$ and $B_{\mathbf{j}\to\mathbf{i}'}$ where $i \neq i'$, admitting change value **j** (in a state $s$) of a cell $a_{x.y,\ldots,z}$ into two distinct values (thus yielding different next states $t \neq t'$). So, multigraph-automata like those in Figs. 5.1 (Chap. 5), 7.1, 7.2 that specify change of cells' value, have to be deterministic. Otherwise one faces a conflict of values on entries in some tables produced in effect of combining c-e structures into cellular ones. These considerations lead to the conclusion:

*The summation of c-e structures is semantically correct if and only if the tables $B_{\mathbf{j}\to\mathbf{i}}$ in the result do not bring about ambiguity in formulas defining behaviour of each cell.*

Finally, let us present an example of a planar cellular structure c-e with hexagonal cells that tessellate the plane, and its evolution starting from a given initial state.

## *Example (Development of Bees)*

Example in Fig. 7.12 shows the evolution of the cellular c-e structure $U^{bee}$, which is a fragment of a honeycomb with cells arranged in a Cartesian plane, as in Fig. 2.9 (Chap. 2). This is a simplified demonstration of the conversion stages of bees from eggs to adult insects. The queen bee is in $a_{00}$ and lays eggs in the surrounding cells. In the next stages of evolution, eggs hatch into larvae, that become converted into pupae, and pupae into adult bees. We omit here some zoological principles of bee nutrition, according to which drones or a new queen bee could be created.

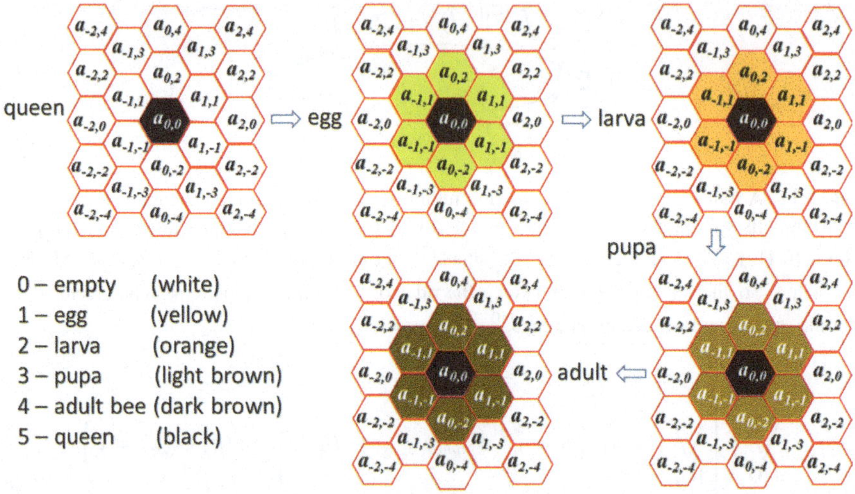


**Fig. 7.12** The evolution of bees from eggs laying by the queen, through larvae and pupae. The initial state $s$ is such that $s(a_{0,0}) = 5$ and remaining cells have value 0

The relation $[\![a_{x,y}]\!]$ which determines behaviour of cell $a_{x,y}$ is defined as: $(s,t) \in [\![a_{x,y}]\!]$ holds if and only if the states $s$ and $t$ satisfy either of the following formulas (using Table 7.4):

$$
\begin{array}{llr}
s(a_{x,y}) = \mathbf{0} \wedge s|_{nbh(a_{x,y})} \in B^{bee} \wedge t(a_{x,y}) = \mathbf{1} & \text{or} & \text{empty} \rightarrow \text{egg} \\
s(a_{x,y}) = \mathbf{1} \wedge s|_{nbh(a_{x,y})} \in B^{bee} \wedge t(a_{x,y}) = \mathbf{2} & \text{or} & \text{egg} \rightarrow \text{larva} \\
s(a_{x,y}) = \mathbf{2} \wedge s|_{nbh(a_{x,y})} \in B^{bee} \wedge t(a_{x,y}) = \mathbf{3} & \text{or} & \text{larva} \rightarrow \text{pupa} \\
s(a_{x,y}) = \mathbf{3} \wedge s|_{nbh(a_{x,y})} \in B^{bee} \wedge t(a_{x,y}) = \mathbf{4} & & \text{pupa} \rightarrow \text{adult}
\end{array}
$$

Conversions other than $\mathbf{0} \rightarrow \mathbf{1}$, $\mathbf{1} \rightarrow \mathbf{2}$, $\mathbf{2} \rightarrow \mathbf{3}$, $\mathbf{3} \rightarrow \mathbf{4}$ do not take place, thus are not taken into account in the semantic table.

Therefore the semantics of the cellular c-e structure $U^{bee}$ is:

$$(s,t) \in [\![U^{bee}]\!] \overset{\text{def}}{\iff} \forall a_{x,y} \in car(U^{bee}) : (s,t) \in [\![a_{x,y}]\!].$$

**End of example.**

**Table 7.4** Integrated semantic table $B^{bee}$, defining functions $f_k^{0\to1}$, $f_k^{1\to2}$, $f_k^{2\to3}$, $f_k^{3\to4}$ ($k = 1, \ldots, 6$) with the same domain $nbh(a_{x,y})$ in the top line and values in the rows underneath. This table, together with the formulas given below it, specifies the evolution of the bee from egg-laying by the queen to the adult insect

| $nbh(a_{x,y})$ | $a_{x-1,y-1}$ | $a_{x-1,y+1}$ | $a_{x,y+2}$ | $a_{x+1,y+1}$ | $a_{x+1,y-1}$ | $a_{x,y-2}$ |
|---|---|---|---|---|---|---|
| egg $f_1^{0\to1}$ | 5 | 0 | 0 | 0 | 0 | 0 |
| egg $f_2^{0\to1}$ | 0 | 5 | 0 | 0 | 0 | 0 |
| egg $f_3^{0\to1}$ | 0 | 0 | 5 | 0 | 0 | 0 |
| egg $f_4^{0\to1}$ | 0 | 0 | 0 | 5 | 0 | 0 |
| egg $f_5^{0\to1}$ | 0 | 0 | 0 | 0 | 5 | 0 |
| egg $f_6^{0\to1}$ | 0 | 0 | 0 | 0 | 0 | 5 |
| larva $f_1^{1\to2}$ | 5 | 1 | 0 | 0 | 0 | 1 |
| larva $f_2^{1\to2}$ | 1 | 5 | 1 | 0 | 0 | 0 |
| larva $f_3^{1\to2}$ | 0 | 1 | 5 | 1 | 0 | 0 |
| larva $f_4^{1\to2}$ | 0 | 0 | 1 | 5 | 1 | 0 |
| larva $f_5^{1\to2}$ | 0 | 0 | 0 | 1 | 5 | 1 |
| larva $f_6^{1\to2}$ | 1 | 0 | 0 | 0 | 1 | 5 |
| pupa $f_1^{2\to3}$ | 5 | 2 | 0 | 0 | 0 | 2 |
| pupa $f_2^{2\to3}$ | 2 | 5 | 2 | 0 | 0 | 0 |
| pupa $f_3^{2\to3}$ | 0 | 2 | 5 | 2 | 0 | 0 |
| pupa $f_4^{2\to3}$ | 0 | 0 | 2 | 5 | 2 | 0 |
| pupa $f_5^{2\to3}$ | 0 | 0 | 0 | 2 | 5 | 2 |
| pupa $f_6^{2\to3}$ | 2 | 0 | 0 | 0 | 2 | 5 |
| adult $f_1^{3\to4}$ | 5 | 3 | 0 | 0 | 0 | 3 |
| adult $f_2^{3\to4}$ | 3 | 5 | 3 | 0 | 0 | 0 |
| adult $f_3^{3\to4}$ | 0 | 3 | 5 | 3 | 0 | 0 |
| adult $f_4^{3\to4}$ | 0 | 0 | 3 | 5 | 3 | 0 |
| adult $f_5^{3\to4}$ | 0 | 0 | 0 | 3 | 5 | 3 |
| adult $f_6^{3\to4}$ | 3 | 0 | 0 | 0 | 3 | 5 |

## References


1. Bramer, M.: Logic Programming with Prolog. Springer (2005)
2. Czaja, L.: Computing in Cause-Effect, Structures. Lecture Notes in Networks and Systems, vol. 331. Springer (2022)
3. Dold, A.: Lectures on Algebraic Topology. Springer (1995); Euler Characteristic: https://en.wikipedia.org/wiki/Euler_characteristic#Plane_graphs
4. Euler, L.: Elementa doctrinae solidorum. Novi Commentarii Academiae Scientiarum Petropolitanae, pp. 109–140 (1758)
5. Gardner, M.: The fantastic combination of John Conway's new solitaire game "life". Sci. Am. **223**, 120–123 (1970)
6. Grunbaum, B., Shepard, G.C.: Tilings and Patterns. Dover Publications Inc., Mineola New York (2016)
7. Hatcher, A.: Algebraic Topology. Cambridge University Press (2002)
8. Mirkowska, G., Salwicki, A.: Algorithmic Logic. Warszawa: PWN-Polish Scientific Publishers, D. Reidel Publishing Company, A Member of the Kluwer Academic Publisher Group, Dordrecht/Boston/Lancaster/Tokyo (1987)
9. Von Neumann, J.: Theory of self-reproducing automata. Published by University of Illinois Press, Urbana, IL, Edited and completed by Arthur W. Burks (1966)
10. Spanier, E.H.: Algebraic Topology. Springer (1982)
11. Hu, S.-T.: Threshold Logic. University of California Press, Berkeley and Los Angeles (1965)
12. Wolfram, S.: Universality and complexity in cellular automata. Phys. D **10**, 1–35 (1984)
13. Wolfram, S.: A New Kind of Science. Wolfram Media Inc., Champaign, IL (2002)
14. Zhang, Z.-Y.: Boolean networks. In: Dubitzky, W., Wolkenhauer, O., Cho, K.H., Yokota, H. (eds.) Encyclopedia of Systems Biology. Springer, New York (2013)

# Index





L. Czaja, *Cellular Cause-Effect Structures*, Studies in Systems, Decision and Control 526,
https://doi.org/10.1007/978-3-031-55590-9

**SPRINGER NATURE**

# GPSR Compliance

*The European Union's (EU) General Product Safety Regulation (GPSR) is a set of rules that requires consumer products to be safe and our obligations to ensure this.*

*If you have any concerns about our products, you can contact us on ProductSafety@springernature.com*

In case Publisher is established outside the EU, the EU authorized representative is:

Springer Nature Customer Service Center GmbH
Europaplatz 3
69115 Heidelberg, Germany

The manufacturer's authorised representative in the EU is Springer Nature Customer Service Centre GmbH, Europaplatz 3, 69115 Heidelberg, Germany. If you have any concerns regarding our products, please contact ProductSafety@springernature.com

Printed and bound by CPI Group (UK) Ltd, Croydon, CR0 4YY
28/09/2026
02224081-0016